Fishing a Borderless Sea

Fishing a Borderless Sea

ENVIRONMENTAL TERRITORIALISM IN THE NORTH ATLANTIC, 1818–1910

BRIAN J. PAYNE

MICHIGAN STATE UNIVERSITY PRESS • *EAST LANSING*

⊚ The paper used in this publication meets the minimum requirements of ANSI/NISO Z39.48-1992 (R 1997) (Permanence of Paper).

Michigan State University Press
East Lansing, Michigan 48823-5245

Printed and bound in the United States of America.

18 17 16 15 14 13 12 11 10   1 2 3 4 5 6 7 8 9 10

LIBRARY OF CONGRESS CATALOGING-IN-PUBLICATION DATA
Payne, Brian J.
Fishing a borderless sea : environmental territorialism in the North Atlantic, 1818–1910 / Brian J. Payne.
p. cm.
Includes bibliographical references and index.
ISBN 978-0-87013-874-4 (paper : alk. paper)
1. Baitfish fisheries—North Atlantic Ocean—History—19th century. 2. Baitfish fisheries—North Atlantic Ocean—History—20th century. 3. Fishers—North Atlantic Ocean—History. 4. Social conflict—North Atlantic Ocean—History. 5. North Atlantic Ocean—Commerce—History. 6. Territorial waters—North Atlantic Ocean—History. 7. Fishery resources—North Atlantic Ocean—History. 8. Fishery law and legislation—History. 9. International economic relations—History—19th century. 10. International economic relations—History—20th century. I. Title.
SH351.B27P39 2010
338.3'72709163409034—dc22
2009027370

Cover design by Erin Kirk New
Book design by Charlie Sharp, Sharp Des!gns, Lansing, Michigan
Cover art: "Baiting Trawls on deck of Gloucester haddock schooner *Mystic*, Captain McKennon." Photograph by T. W. Smillie. Image ID: fig0049, NOAA's Historic Fisheries Collection. NOAA National Marine Fisheries Service. http://www.photolib.noaa.gov/brs/hfind1.htm

green press INITIATIVE Michigan State University Press is a member of the Green Press Initiative and is committed to developing and encouraging ecologically responsible publishing practices. For more information about the Green Press Initiative and the use of recycled paper in book publishing, please visit *www.greenpressinitiative.org*.

Visit Michigan State University Press on the World Wide Web at *www.msupress.msu.edu*

Contents

Acknowledgments

This book is the product of over ten years of research, writing, and editing. Throughout this decade I received kind advice from mentors, colleagues, and friends. Recently I have benefited from the sound council of Michael Chiarappa, Jonathan Phillips, and Michael Carhart. I thank Julie Loehr and Kristine Blakeslee, at Michigan State University Press, and Barbara Fitch Cobb for their patience and editing skills.

There are many teachers and mentors that I need to thank. Drs. Donald Bain, Stephen Valone, and M. Biskumpski guided me through my undergraduate education at St. John Fisher College. In graduate school I found many excellent mentors in Marli Weiner, Liam Riordan, Stephen Miller, Stephen Hornsby, Jacque Ferland, and Richard Judd, but I am especially indebted to Scott See, my advisor, mentor, and friend. Scott patiently guided me through many drafts and has written more letters of recommendation that any person should rightly request.

Throughout this project I received generous financial support from the Department of History and the Canadian-American Center at the University of Maine. A grant from the Massachusetts Historical Society provided me additional funding to research at the Maine Historical Society, the Baker Library and the Langdell Library at Harvard University, and the Mystic Seaport Museum and Archives. A fellowship from the Canadian Embassy in Washington, DC, provided funds for a lengthy research trip to the Nova Scotia Archives and Records Management Office in Halifax. A faculty research grant from Old Dominion University provided funds for one last research trip to Nova Scotia and to the Public Archives and Records Office of Prince Edward Island. I would also like to thank the kind people at the Gorsebrook Institute at St. Mary's University in Halifax, Nova Scotia for their hospitality and assistance.

More important than these professional thanks, I owe a great debt of gratitude to my friends and family for their support and valuable distractions. I must close these acknowledgements with a thank you to the most important person in my life; my wife Shandra, who never lost faith in my ability even when I was certain I had failed.

I dedicate this book to our child, who will arrive shortly to bring new meaning to our lives.

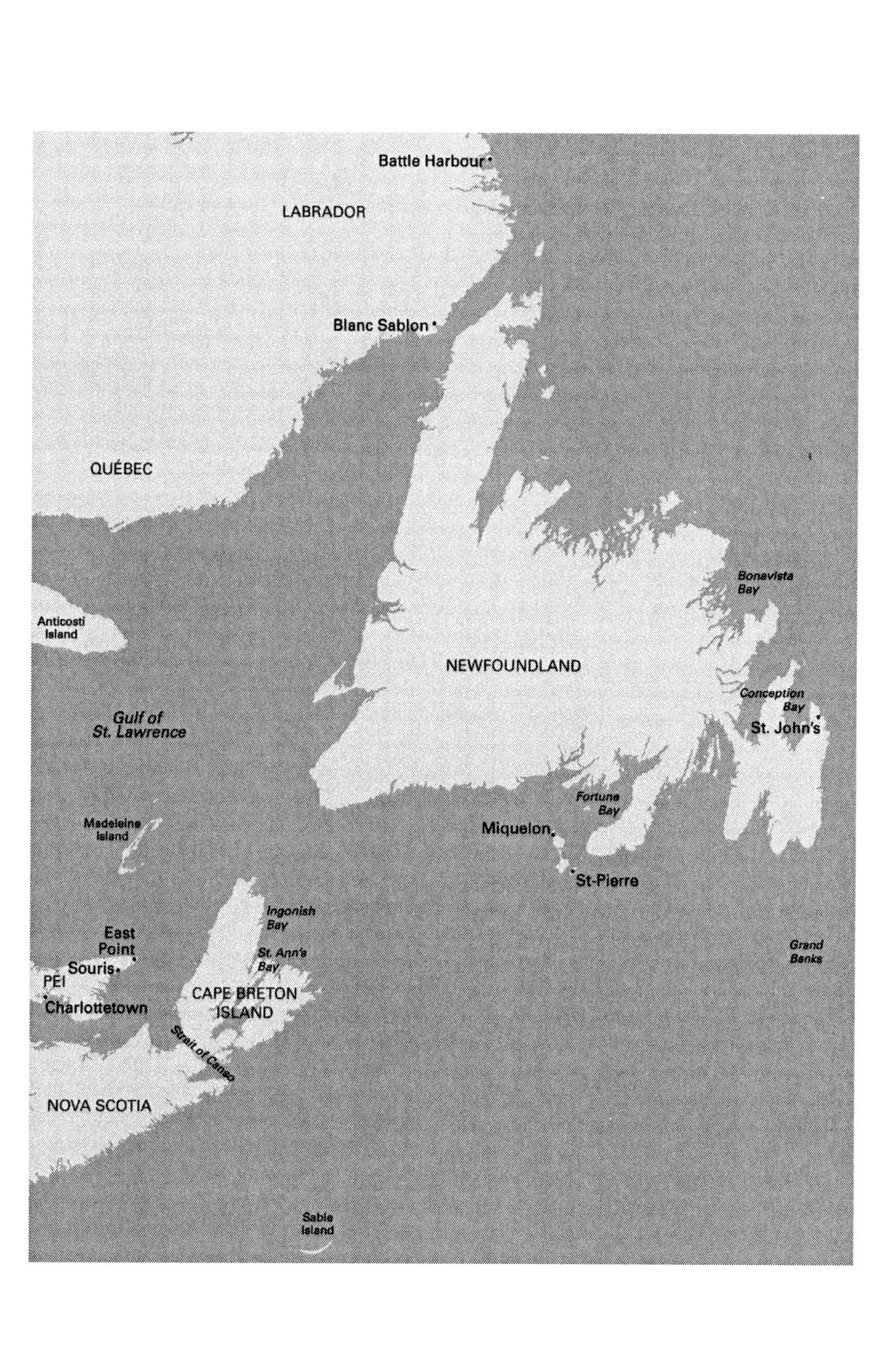
Battle Harbour
LABRADOR
Blanc Sablon
QUÉBEC
Anticosti Island
Gulf of St. Lawrence
NEWFOUNDLAND
Bonavista Bay
Conception Bay
St. John's
Fortune Bay
Miquelon
St-Pierre
Madeleine Island
Ingonish Bay
St. Ann's Bay
East Point
Souris
PEI
Charlottetown
CAPE BRETON ISLAND
Strait of Canso
NOVA SCOTIA
Grand Banks
Sable Island

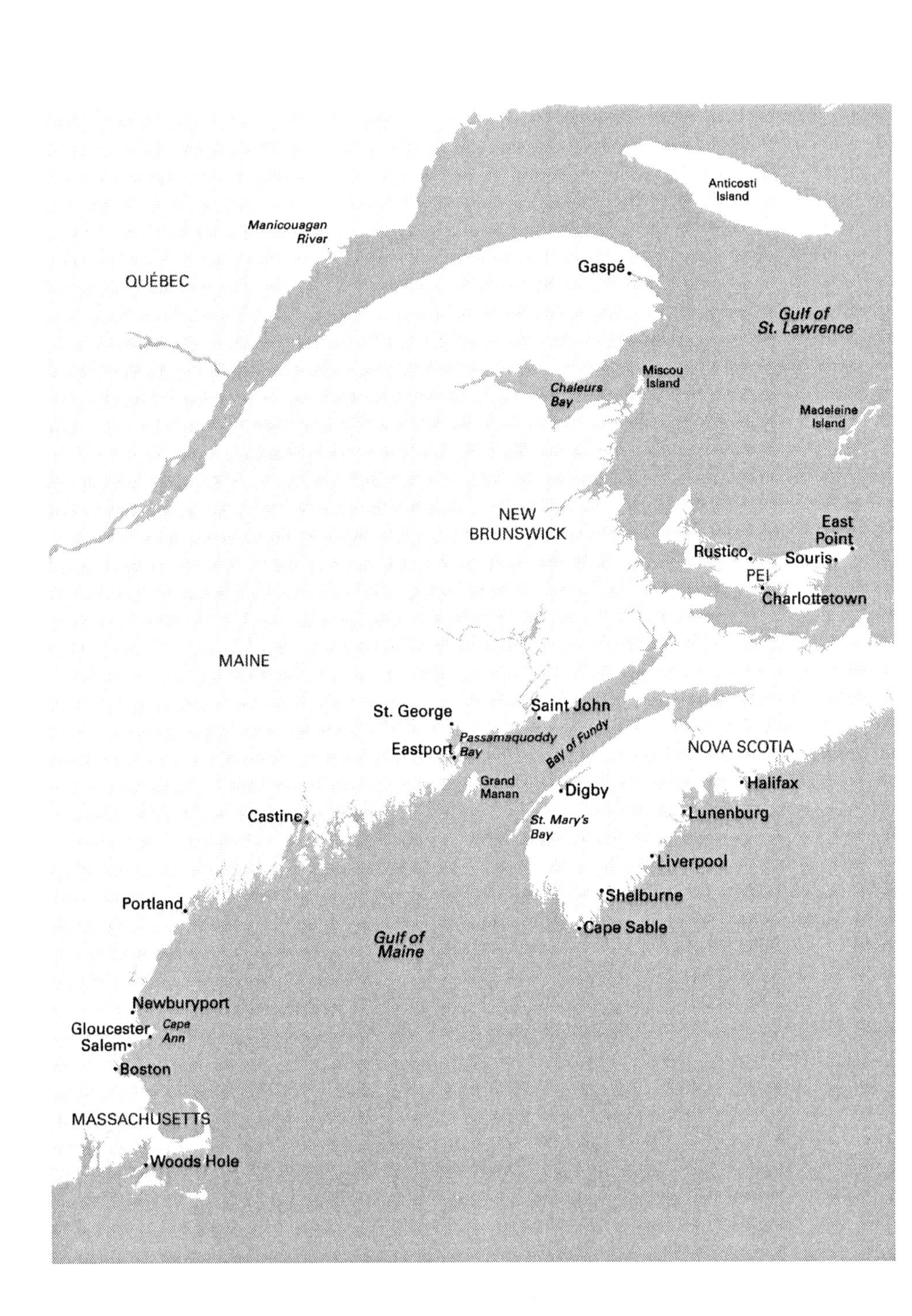
Anticosti Island
Manicouagan River
QUÉBEC
Gaspé
Gulf of St. Lawrence
Miscou Island
Chaleurs Bay
Madeleine Island
NEW BRUNSWICK
East Point
Rustico
Souris
PEI
Charlottetown
MAINE
St. George
Saint John
Passamaquoddy Bay
Bay of Fundy
NOVA SCOTIA
Eastport
Grand Manan
Digby
Halifax
Castine
St. Mary's Bay
Lunenburg
Liverpool
Shelburne
Portland
Cape Sable
Gulf of Maine
Newburyport
Gloucester
Cape Ann
Salem
Boston
MASSACHUSETTS
Woods Hole

Introduction

We must have bait or quit fishing.
—Alden Kinney, master of the *David J. Adams*

MANY HISTORIANS HAVE examined the various and complex histories of the oceans' great fisheries. Within this larger fisheries history, the North Atlantic has been one of the primary geographies of historical investigation. The North Atlantic codfish has captured the imagination of writers for hundreds of years. The mighty codfish emerged as one of the most important commodity trades in the Atlantic world and built many of the great merchant houses of England.[1] No doubt, the cod reigned supreme in human exploitative efforts to gain wealth and protein from the vastness of the oceans. But in all our efforts to examine the history of the great cod fisheries, we often forget some of the peripheral economies that developed to supply, process, ship, market, and sell the finished product to the hungry consumers. Bait was the foundation of the cod fisheries, and by the nineteenth century, fishermen were paying top dollar for good, fresh bait. Long before scientists and politicians recognized the overexploitation of codfish, fishermen searched near and far for bait. After a thorough examination of the historical record, it becomes clear that baitfish supplies ran low long before codfish supplies. It was a simple matter of numbers. For every codfish caught, North Atlantic fishermen put hundreds of baitfish on hooks and strung them out across the ocean. According to the 1887 report by George Brown Goode, the average American fishing schooner working the banks of the North Atlantic carried 150 to 200 barrels of baitfish per season, sometimes using as much as 250 to 300 barrels.[2] As a result, baitfish became scarce, and with scarcity came controversy. For all the talk about cod, it was actually the humble herring that stirred up most of the international controversy and labor violence in the North Atlantic.

This book is not about the mighty codfish, but instead seeks to explain the dependency on, complexities of, and controversies over baitfish in the North Atlantic throughout the nineteenth century and into the early twentieth century when non-bait fishing, principally trawling, restructured the entire operation. This history carries with it an interesting and unique dynamic. By the nineteenth century, bait fishing and cod fishing were two very different occupations. Bait fishing was largely the work of small-scale, localized weir fishermen, upon whom the more capital-intensive fishing operations of places like Gloucester and Lunenburg depended. Deep-sea fishermen depended upon inshore fishermen for baitfish as much as inshore fishermen depended upon deep-sea fishermen for a ready market for their own specialized commodity. This interdependency was well established by the nineteenth century as large-scale operations increasingly specialized in the capture of the mighty codfish and thus outsourced the bait fishery to smaller, localized operations. This mutual dependency made both compromise and conflict inevitable.[3]

This relationship carried with it yet another geographic characteristic. Deep-sea fishermen desperately wanted *fresh* bait, and before the introduction of shipboard icing facilities, fresh bait had to be local bait. Gloucester fishermen could only carry so much baitfish with them to the fishing grounds before it spoiled—after which, they depended upon local fishermen in Canada and Newfoundland to supply them with the necessary baitfish. This gave local fishing communities bordering the North Atlantic fishing grounds the opportunity to control a market niche, set extraction levels, establish pricing, and control access to their meager, yet essential, local baitfish grounds. According to George Brown Goode, American schooners had to travel to Canadian harbors three or four times each season to acquire fresh bait from local fishermen. These "baiting trips" occupied the American fishermen anywhere between four days to three weeks each, and thus distracted the American efforts away from the more profitable cod fisheries. Goode went so far as to claim that "One of the greatest needs of this fishery is the invention of some method by which the necessary supply of bait can be obtained by the bankers without interruption of this kind."[4]

In order to establish authority over their immediate environment, local fishermen constructed what might be understood as "informal codes of conduct."[5] These informal codes of conduct were rules created by the local community that did not necessarily correspond to official law and in fact often directly contradicted that law. In the North Atlantic bait fishery, the primary informal code of conduct was the exclusion of all "outsiders" or "strangers" from the actual act of catching the fish. Visitors could, and often did, enter inshore waters to purchase baitfish from the local fishermen, but these local fishermen often responded with crowd action, intimidation, and violence whenever any outsider attempted to catch their own baitfish in the local environment. This was true even in those circumstances when the outsiders had the right, by law or by treaty, to fish in local inshore waters. Furthermore, "outsider" was not a category exclusively reserved for foreign fishermen. An outsider was any fisherman or fishing vessel that the local community so defined and often included individuals hailing from the same state, province, country, or empire as that local community. It is essential to note that while many Americans found themselves in conflict with Canadians, the whole history of the controversy cannot be defined as strictly an American-Canadian conflict.

In addition to this principal informal code of conduct restricting the catching of fish to local fishermen, there was a secondary rule that helped structure the bait fishery. Outsiders were normally permitted to purchase baitfish regardless of formal law or treaty rights that may have stipulated otherwise. At numerous times throughout the history of the North Atlantic fisheries, local and national governing bodies attempted to outlaw the selling of locally caught fish to foreign fishermen. Despite these laws, local fishermen often continued operating their business under the guiding principals of the accepted informal codes of conduct, which allowed for such trade so long as those foreigners did not attempt to catch their own. In many cases, the informal codes of conduct overruled formal law and treaty stipulations. The combined influence of these two informal codes of conduct enabled local fishermen to retain territorial control over their fishery resources while still permitting them to participate in the global fishery economy.

The construction, interpretation, and enforcement of these two primary informal codes of conduct represent one level of territorialism in the North Atlantic. The ability of local fishermen to enforce the codes defined the level of their territorial success. Formal law did work its way into the fishery as a second level of territorialism. The principal legal construct in this formal territorialism was the Convention of 1818 signed between the United States and Great Britain shortly after the conclusion of the War of 1812. The Convention of 1818 was the standing rule related to the rights and restrictions of American fishermen seeking to utilize the resources of inshore waters in the British colonies, and then the Canadian provinces, in the northwest Atlantic and was thus, unlike the informal codes of conduct, specifically drawn along national lines. The primary controversial clause within that agreement simply stated that "the United States hereby renounces forever, any Liberty heretofore enjoyed or claimed by the Inhabitants thereof, to take, dry, or cure Fish on, or within three marine Miles of any of the Coasts, Bays, Creeks, or Harbours of His Britannic Majesty's Dominions in America," while providing American fishermen the continued luxury of entering British ports "for the purpose of Shelter and of repairing Damages therein, of purchasing Wood, and of obtaining Water, and for no other purpose whatever."[6]

Nowhere is bait specifically referred to in the Convention of 1818, but it nonetheless became the principal concern among fishermen, politicians, and diplomats throughout the nineteenth century as they struggled with what became known simply as the "fishery question." The "question" was: can American fishermen enter British and/or Canadian waters and harbors to hire crew, transship cargoes, or to purchase supplies, principally bait, without violating the terms of the 1818 agreement? The United States government continually argued that the Convention of 1818 was intended to regulate the fisheries only, and not the trade of commodities or ancillary economic activities related to fishing. As such, American fishing schooners could be restricted from fishing or preparing to fish within territorial waters, but American fishermen should enjoy the same rights of free trade in ports and harbors as other American traders. The Canadian government, which was only occasionally supported by British authorities, claimed that the Convention of 1818 was intended to protect the unique advantages of the British and Canadian possessions in the North Atlantic, and if American fishermen wanted equal access to Canadian waters, then the United States government must provide Canadian merchants equal access to the American market. Essentially, Canadian representatives throughout the period called for fair trade rather than simply free trade. The Canadian case rested heavily on the "for no other purpose whatever" clause of the 1818 agreement that restricted all activities of American fishermen in Canadian ports other than shelter, repair, wood, and water.

This question was complex, and the vague wording of the Convention of 1818 did not help. Therefore, diplomats often sought to write new language, always temporary in duration, to more clearly define the rights and limits of American fishermen. Because the fish trade, including baitfish, was so closely related to more general trade policies, the specifics of this debate often became just one part of more general free-trade politics in the United States, Great Britain, and the Dominion of Canada. The Reciprocity Treaty of 1854, which lasted until 1866, and the Treaty of Washington of 1871, which lasted until 1885, were two momentary pauses in the great diplomatic debate over bait, because the formality of law and treaty largely recognized the informality of the codes of conduct that permitted the free trade of baitfish between British and American fishermen in the northwest Atlantic. Controversy, both between fishermen and between government officials, largely erupted only when the governments sought to impose formal regulations that did not correspond with the informal codes of conduct.

This history is not one that is exclusively limited to diplomacy or politics. Through the examination of the historical record, one is often surprised at the amount of control or authority local fishermen had in the handling of this massive geopolitical debate. By the early nineteenth century, Canadian and British fishermen often took employment aboard American fishing vessels and thus naturally facilitated increased economic cooperation between those Yankee schooners and the local fishing communities. Many records indicate that a sizable share of the American fishing work force during the nineteenth century were British subjects of Canada, and in many cases even the masters of the Yankee schooners were Canadian-born. This employment pattern enabled British and Canadian crew members and masters to negotiate informal trade contracts with local communities beyond the eyes of customs officials. During the two periods of free trade and free fishing, 1854–1866 and 1871–1885, this commerce was officially deemed as legal trade, but beyond those periods, the various governments considered the trade illicit acts of smuggling. While governments, from time to time, may have changed the terminology related to the bait business from trade to smuggling, fishermen changed few of their actual habits. Despite the illegality of the bait trade in 1870 or 1886, American fishermen still found ready sellers of bait, ice, and other supplies in Canadian and British ports. The commerce was nearly impossible to regulate, and the situation got so bad that by the twentieth century, the Newfoundland government went so far as to outlaw the employment of local fishermen aboard American schooners in a failed effort to undercut the Americans' primary source of negotiating power.

During the active life of the Convention of 1818, which lasted from 1818 to 1854; 1866 to 1871; and 1886 to 1910, fishermen worked out a difficult balance between the informal codes of conduct created by the communities and the formality of law and treaty. At times these came into direct collision as fishermen attempted to enforce or restructure the balance between the two levels of territorialism. The result of this collision was often violence, vessel seizures, or diplomatic discourse. These moments in history, ones that generated public, political, and/or diplomatic publicity, offer historians a rare opportunity to read the thoughts of a largely illiterate working class of fishermen who practiced their craft in a large and contested environment. From this examination, we learn that the North Atlantic fisheries were seldom peaceful. Yet, the story of conflict that emerges is not one that would normally be expected. Instead of geopolitical conflict between nationally defined fishing groups and their political representatives, we uncover a much more complex pattern of social violence among the fishing laborers. In the North Atlantic, fishermen constructed a host of informal codes of conduct, often outside the parameters of formal law, as a method to regulate what they often perceived to be locally defined resources. Local fishermen did not universally exclude outsiders from their local environments, but instead only demanded that visiting fishermen conduct themselves in accordance with these locally defined codes of conduct. Cooperation and compromise resulted from the adherence to local codes. Conflict emerged only when visiting fishermen, foreign or domestic, ignored local codes of conduct. For the fishermen in the North Atlantic, formal law and diplomacy did not matter nearly as much as community codes of conduct, which, to a large extent, regulated access to local resources throughout the nineteenth century and were only replaced by more formal law once diplomats sought to establish international regulation in the North Atlantic.

This work attempts to place the labor history of the North Atlantic bait trade within a larger diplomatic context. To do so, it is best to understand the value of looking beyond strict geopolitical divisions in structuring the study. This work is largely an effort to understand commerce and labor in the North Atlantic borderlands. Regional identity is difficult to establish for individuals who work at sea. There are many cases in which the fishermen of a nation defined the border in a different manner than did their political leaders. Therefore, in order to understand the dynamics of the fishing industry, one must include international economic trends and not restrict the study of this transnational industry to the current nations, provinces, or states. Borderlands theory provides a medium in which North American historians can explore these international or interregional themes.

Garth Stevenson defined a borderland region as "a territorial entity having some natural and organic unity or community of interest that is independent of political administrative boundaries," or as "a 'natural' regionalism that contrasts with their own artificiality."[7] This statement leads to an understanding of regionalism with little to no reference to international border making. Similarly, American historian C. Vann Woodward argued: "To limit the subject of historical study within national boundaries is always to invite the charge of narrow perspective and historical nationalism."[8] This borderland approach to North American history can greatly enrich a study on territorial disputes in the North Atlantic fishery.

This work defines a borderland—or its maritime translation, a border-sea—as an area in which a politically constructed border divided a region that would have otherwise shared some communal sense of identity. This unity was often the result of similar geographic, social, religious, and/or economic structures. In his study of Canadian regionalism, Stevenson noted that "an analysis of Canadian regionalism that fails to view it in its North American context is incomplete and seriously misleading."[9] Therefore, the history of the North Atlantic fishing industries in the Atlantic Provinces and New England cannot be thoroughly examined without an understanding of the borderlands aspect of this specific region. An industry that was so heavily dominated by international trends requires the incorporation of the borderland philosophy that has been developed by geographers such as Victor Konrad, who argued:

> The borderlands of the United States and Canada are distinctive regions of mitigating landscapes fading from the common edge of the boundary. Borderlands evolve from border regions, or contiguous zones in which exchange between nation-states takes place. . . . In the Canada–United States context particularly, borderlands are regions that have a tempering

> effect on the central tendencies of each society, and these regions reveal the ways in which the nation-states blend into each other.[10]

This work argues that the North Atlantic fishery was such a "contiguous zone" that facilitated international contact. Both maritime and environmental history can greatly benefit from looking beyond geopolitical borders that may artificially separate regions or environments.

Politicians and diplomats worked hard throughout the nineteenth century to create law and order in the North Atlantic fisheries according to national boundaries. There was a constant effort to define the limits of territorial authority, but fishermen also played a part in this construction by challenging those limits. Fishermen created unique identities that more often reflected their own understandings of the environment as laborers and resource commodities traders, rather than as members of any legally defined nation-state. This history of the bait fishery and trade thus seeks to understand how the laborers themselves constructed resource extraction policy based on their own ideas of environmental stewardship. Environmental history, however, does not often give much credit to working people in capitalist economies. Marxist philosophies and structures have had a profound influence over the historians' interpretation of human use of natural products. These Marxist interpretations paint a stark picture of capitalism and its conquest over nature. Donald Worster's seminal book *Dust Bowl: The Southern Plains in the 1930s* argued that this ecological disaster on the American prairies had more to do with farmers' economics than with nature.[11] Early in his analysis, Worster stated that "The Dust Bowl . . . was the inevitable outcome of a culture that deliberately, self-consciously, set itself that task of dominating and exploiting the land for all it was worth."[12] Worster showed how American farmers sought to reconstruct the western plains into an agricultural system in tune with the capitalist needs of the country, and thus he argued that capitalism dictated Americans' use of natural resources. This economic and cultural idea led to the desire for total use of all resources as rapidly as possible, regardless of the ecological situation. Worster studied farmers within an industrial context. These farmers labored in a capitalist economy just like the industrial workers of the East. He stated:

> Capitalism . . . has been the decisive factor in this nation's use of nature. To understand that use more fully we must explain how and why the

> Dust Bowl happened, just as we have analyzed our financial industrial development in the light of the 1929 stock market crash and the ensuing shutdowns.[13]

While his work is truly insightful, this interpretation of resource use limited Worster's ability to see any personal attachment the farmers had with the land. He did not incorporate the farmers' desire to preserve the culture of their resource industry. Resource users, whether they were farmers or fishermen, did not work in isolation from the environment in which those resources existed.

Theodore Steinberg added the theories of industrialization to environmental history in his chronicle of river management in New England. His book *Nature Incorporated: Industrialization and the Waters of New England* concentrated on the ideas of ownership of natural resources and the quest to manipulate these resources in order to provide for the emerging industrial order.[14] He argued that for the industrialists, the river and the laborers were two sides of the same coin. Both were sources of work (energy) that should be controlled, restricted, and structured according to industrial needs and norms. Unlike Worster, Steinberg did explain some signs of resistance to this industrial capitalist order, but this movement came from local farmers who sought to capitalize the water and utilize it for their own means. Thus, according to Steinberg, they also saw nature in strict terms of wealth accumulation.

These interpretations of human/nature relations have not remained unchallenged. More layered interpretations of the history of resource use have included noneconomic factors and have shown that Americans are not always limited by a strict interpretation of capitalism. Bonnie McCay and James Acheson challenged the "tragedy of the commons" thesis in their edited collection of essays, *The Question of the Commons: The Culture and Ecology of Common Resources.*[15] Garrett Hardin's thesis on the tragedy of the commons claimed that in common-property situations, resources would always be immediately exploited to maximum personal profit. Hardin argued that in order to avoid this situation in a capitalist society, the resources must either be privatized or placed under strict governmental regulation.[16] Hardin assumed that those working in the commons had no desire to preserve their resources or to protect the common property. McCay and Acheson argued that the tragedy-of-the-commons thesis reduced resource management and environmental problems to issues of property rights and ignored more complex socioeconomic systems or cultural constructions related to noneconomic value of nature and natural resources. Their critique argued for a deeper understanding of resource use that went beyond a Marxist analysis by including the cultural-based ethics of preserving the common resources.

Much of James Acheson's subsequent work focused specifically on local stewardship in resource management in Maine's lobster industry. His latest book, *Capturing the Commons: Devising Institutions to Manage the Maine Lobster Industry*, introduced the key concept upon which this work is based—that is, the idea of what he called formal and informal codes of conduct.[17] Although his work primarily addressed public-policy concerns rather than historical dynamics, his theories regarding resource users' propensity to set limits upon themselves (informal codes of conduct), often more strict than those set by state or federal governments (formal codes of conduct), illustrate the great flaw in the Marxist approach to environmental history. Acheson referred to the ways in which the state and federal governments and the fishing population itself cooperated in establishing limitations on productivity. Fishermen conducted this form of territorial regulation. Acheson stated that the Maine lobster industry is a case "where local-level communities and governments have been able to generate rules to effectively manage resources at sustainable levels."[18] This self-regulation of specifically assigned territories ultimately led to conservationist efforts. According to Acheson, local resources policed by local producers leads to sophisticated stewardship. He argued, "In retrospect, those concerned with the lobster fishery have worked hard to maintain the fishery for themselves and future generations. To this end, they have developed several different kinds of rules to limit access to the resource and to control the fishery, a common-pool resource. They are truly 'capturing the commons.'"[19] In this sense, the resource users, the lobstermen themselves, were certainly more than just industrial workers who could not control the means of production because of the forces of industrial capitalism. Through historical investigation, we find the same to be true among small-scale bait fishermen in the North Atlantic.

Historian Richard Judd further challenged the Marxist critique of land use policies in two important works. In his little known and significantly underappreciated work *Aroostook: A Century of Logging in Northern Maine*, Judd introduced the idea of local desire to preserve forest resources in Aroostook County, Maine.[20] This desire was closely associated with the cultural identity the inhabitants created around the existence of the forest and the lumber industry. This cultural identity set up a framework upon which resource users could create conservationist ideologies. According to Judd, Maine lumbermen were not simply capitalists or industrial workers who viewed the resource in strict terms of dollars and cents. Clearly something more complex than simply supply-and-demand economics shaped their ideas regarding conservation.

In *Common Lands, Common People: The Origins of Conservation in Northern New England*, Richard Judd showed how this environmental ethics existed throughout New England. This seminal work, perhaps more than any other, illustrated the complex cultural and social ideologies that formed the backbone of how individuals chose to use nature for economic ends. Judd challenged "environmental historians to look more closely at the people who used these resources and, in the second half of the nineteenth century, pondered their conservation."[21] Unlike the political actors in state and federal capitals, locals created a moral value system in regards to their land-use ethics. This moral vision consisted of a belief in common stewardship and communal respect for the balance within nature and the balance between nature and man's use of resources. Judd's book suggested that local inhabitants, who held a deeper understanding of nature and how it can be used and preserved simultaneously, led the way in conservationist efforts by the end of the nineteenth century. He found that local New England farmers, lumbermen, and fishermen built up their own "common stewardship" and discovered that they had

> developed powerful attachments to a familiar landscape and to the natural dynamics that sustained it. This popular ethic was neither uniformly conservationist nor anticonservationist, as we define these terms today, but it was indeed a force to be reckoned with. . . . It inspired a penetrating search for the regularities and harmonies of nature, and it gave local land-use practices a definably moral cast.[22]

The morality involved in conservationism and environmental stewardship, as presented by Richard Judd, along with such borderland studies as that of Victor Conrad, and the debate concerning maritime common property vocalized best by Jim Acheson and Bonnie McCay, constitute the principal theoretical backdrop for this work on the bait fisheries.

The first chapter of this work illustrates why and how British Canadian fishermen sought trade with, and eventually employment aboard, American schooners. The American bounty system and the growing urban market for fish products in the United States, in conjunction with the gradual dismantling of the British mercantile system and the lack of government support in Nova Scotia, set the economic context for a mass out-migration of fishing labor from Nova Scotia. This migration provided the Yankee fleet not only badly needed skilled labor, but also an avenue through which they could conduct trade with local fishing communities in the British Atlantic, which became one of the most important and necessary components of American fishing efforts in the North Atlantic. Historical evidence clearly illustrates that by the 1830s, such trade had become a common practice despite both domestic law and international treaty that prohibited such action. The Reciprocity Treaty of 1854 temporarily legalized this bait smuggling, but when the United States repealed the agreement in 1866, the business was again illegal.

By the late 1860s, the Dominion of Canada emerged as a semi-independent nation within the growing British Commonwealth. Chapter 2 examines how pro-Confederation politicians in Ottawa and Halifax saw the illegal bait trade and the lack of free trade between Canada and the United States as important political issues that could be used to win Nova Scotia support for both Confederation and eventually the National Policy. If these Canadian politicians could convince skeptics in Nova Scotia, as well as New Brunswick and eventually Prince Edward Island, that they could better look after their precious fishery economy than London, then they could more fully integrate them into the Canadian national economic and political systems. For these politicians, the best way to incorporate the Atlantic fisheries into this national agenda was to shut down this illegal bait trade and thus evict all American interests from the Atlantic Provinces. Between the repeal of the 1854 treaty in 1866 and the signing of a new treaty in 1871, Canada sought to impose national political ideas and legislation upon the local informal codes of conduct so as to further separate Nova Scotia fishermen from their Yankee counterparts. To a large extent this effort failed, and local fishermen from around the northwest Atlantic continued to operate according to their informal codes of conduct despite new law or treaty stipulations.

The 1871 treaty was also short-lived, and the United States repealed the agreement in 1885. Chapter 3 examines how Canada again sought to impose national legislation on the bait trade in Nova Scotia, New Brunswick, and Prince Edward Island after this repeal. Despite their effort to prohibit American inshore activities, Yankee fishermen had little problem finding ready sellers of bait and ice supplies. This was largely due to the fact that the vast majority of the crew members, as well as the masters, of American schooners were themselves Canadian or Canadian-born. After Canada utilized a number of high-profile seizures to force the United States into new treaty negotiations that would win them free markets in the United States, the situation grew into an intense diplomatic debate between the United States, Canada, and Great Britain. Efforts to sign a new treaty failed in 1888, and the Canadian government abandoned their efforts to impose new national policy upon the local fishing trade. The bait trade in the northwest Atlantic thus remained under the authority of local codes of conduct.

Beginning in the late nineteenth century, international law emerged as a new force in diplomacy. Leading advocates for the power of arbitration took the fishery question to The Hague Court in 1910. Chapter 4 examines how lawyers attempted to use legal phraseology to restructure authority in the North Atlantic bait fishery—this time the debate was largely focused on Newfoundland. Lawyers at The Hague cited an immense amount of legal precedents to argue differing perspectives as to the rights of Americans in the inshore waters of British possessions in the Atlantic. By the end of the debate, the decision of the court largely formalized the preexisting informal codes of conduct. By stating that local communities could control the specifics of the fisheries so long as Americans were permitted some access, the court merely put into formal words informal codes of conduct that had been at work for nearly a hundred years.

The real blow to local control of inshore bait fisheries had little to do with national law or international agreement. The conclusion of this work points to the effects of both the rising sardine canning industry, which redirected efforts of herring fishermen away from bait production to sardine production, as well as the rise of non-bait fishing that together largely ended local control of inshore herring fisheries. These topics, however, are well beyond the limits of this work and are therefore left to another date or another historian.

Throughout the nineteenth century, the principal issue in the relationship between fishing groups of the North Atlantic was the access to, and control of, local bait supplies. At the same time, the geopolitical forces of the United States, Great Britain, Canada, and the British colonies of the North Atlantic also focused their diplomatic and political efforts on the contentious bait trade. Thus, the great debates over North Atlantic fisheries during the nineteenth century were largely confined to the inshore waters from the Bay of Fundy to the Gulf of St. Lawrence and the seemingly hundreds of small coves and inlets in which bait was caught. Although the geography of the North Atlantic at first appears vast, this work largely focuses only on the inshore waters in which the trade in baitfish occurred. It was in this geography that local communities sought to take control of the rules of extraction and trade, and it was this geography that became the topic of heated international dispute. The work only occasionally ventures out onto the great Grand Banks, Georges Banks, or other offshore fishing environments.

The diplomatic records related to the fisheries dispute largely focus on the geopolitical control of the inshore waters. Yet, within these political and diplomatic records, we often find the words of the average fishermen. By reading these words through new lenses, instead of simply relying on the contemporary political and diplomatic analysis, we can learn much about compromise, cooperation, and conflict within the working environment of the North Atlantic fisheries. By doing so, we uncover a complex system of locally defined codes of conduct that regulated resource extraction in inshore waters. These codes of conduct sought to limit access to resources so as to corner the market for the local community, and thus naturally limited total extraction from the environment to the capacity of those small-scale, local fishermen who often shunned more exploitive technologies such as the purse-seine net or the bottom trawler. What we uncover is an interesting and insightful history of a locally defined environmental stewardship practiced by the very people responsible for resource extraction. What we see in this history of the bait fishery is a local or regional response to the growing influence of internationalism, and perhaps historians, who today are so heavily invested in international and global perspective, should return to the local level in an effort to understand the very intimate connections between working people and their environment.

"White-Washed Yankees"

THROUGHOUT THE FISHING season of 1836, Liverpool (Nova Scotia) fish merchant Philip Carten traveled through the small fishing villages of western Nova Scotia. Like other Nova Scotia fish merchants, Carten sought out inshore fishermen from whom he could purchase baitfish, principally herring or mackerel, in order to outfit his offshore fishing vessels for a voyage to the Grand Banks off Newfoundland. This offshore fishery depended upon inshore fishermen to supply the bait needed for their voyages, thereby allowing them to concentrate their efforts strictly on the catching of ground fish such as cod and halibut. By 1836 this system had been the standard practice for at least a generation. Like the vast majority of Nova Scotia fish merchants, Carten purchased fish via company credit; instead of paying cash for his supplies, he sought to extend store credit to the inshore fishermen. This season, however, Carten was unable to secure any baitfish from the numerous fishermen scattered across the coast of western Nova Scotia. They instead preferred to wait for the arrival of New England fishing schooners than to sell their fish to domestic merchants.

In a letter to the Assembly of Nova Scotia, Carten complained bitterly that this trade practice threatened the complete collapse of Nova Scotia's fishing industry. American trade challenged the historical relationship between merchants and fishermen in Nova Scotia, he argued, and the entire domestic fishing industry operated upon the basis of this system of credit and debt. Carten charged that the American vessels arrived in the harbors of Nova Scotia "having onboard Gin, Boots, and Shoes, Apples, Soap, and other articles and open a regular Trade with the Fishermen and sold the above Goods, taking in return Mackerel."[1]

Although merchants like Philip Carten considered this illicit bait trade as disloyal to their native land and industry, the extent to which it operated clearly shows that the fishermen themselves found it quite lucrative. Carten confronted the disloyalty of native fishermen and the legality of the presence of the foreigners. In his letter to the House of Assembly, he expressed his frustrations and stated that many in his class felt "indignant at the preference given to Foreigners told them they had no business there. . . . stated to the people that they were injuring themselves and robbing the Country of its living."[2] Yet the introduction of American goods and capital offered an alternative to the monopoly held by Nova Scotia merchants and provided another source of income for the small-scale fishermen of the province.

The close cooperation between American fishing schooners and local Nova Scotia bait fishermen that developed during the first half of the nineteenth century challenged the economic control of Nova Scotia fish merchants by introducing American capital into the domestic economy of Nova Scotia. This began immediately following the American Revolution with the arrival of large-scale fishing schooners from Gloucester into the inshore fishing areas of Nova Scotia. By the 1830s, it developed into extensive networks of trade and smuggling and expanded to the direct employment of Nova Scotia fishermen by New England's fishing fleet. These "White-Washed Yankees," as they were often labeled, used the closer cooperation with American fishing interests as a means of economic independence, which illustrates their understanding and use of larger global networks within the North Atlantic fisheries. Local fishing communities could then utilize these global networks in their efforts to gain increased control over their most immediate environment and resources—principally inshore waters and baitfish.

The increased productivity that these American schooners brought to the North Atlantic resulted from a rapid industrialization of New England's fishing fleet. This new industrial fishery, centered on the ethics of American capitalism, which Philip Carten and his fellow Nova Scotia merchants viewed as distinctive from English mercantilism, brought added competition for English fish merchants in the form of fish production, global marketing, and labor recruitment. The industrial fisheries of New England sought to concentrate their efforts on the profitable market fish, like cod, while outsourcing the supportive economies, like bait fishing, to smaller operations. This new dynamic in the fisheries forced a reevaluation of local resource use and environmental management techniques, and all of these factors played a role in the dramatic and sudden shift in economic power over the North Atlantic fisheries. The concentration of power in New England fisheries resulted from unique economic and political developments that were not replicated in Atlantic Canada. This New England fishing economy benefited from cash payment to workers, government bounties, and a rapidly expanding protected domestic market. These features aided in the recruitment of labor and facilitated trade beyond United States territorial limits.

Two strikingly different methods of payment existed for fishing laborers during the nineteenth century. The first system utilized a method of debt-credit relationship that tied merchants and fishermen together—often referred to by historians and economists as either "truck" or "clientage." The second system incorporated a currency or wage-labor method based on work time or production output. The latter relied on access to liquid capital, as well as a substantial local labor force, and therefore only industries that existed in a major metropolis, such as Boston or Gloucester, Massachusetts, could adopt such a system. Conversely, the truck system was most successful in peripheral areas with limited capital and a small labor force.[3] The arrival of an American fleet that practiced wage labor presented significant problems to the merchants of Nova Scotia because it threatened the economic bond between the employers and the workers, and therefore challenged their socioeconomic control over the labor force by presenting the fishermen with another viable option. For this reason, the merchants of Nova Scotia attempted to strengthen the truck system by pressing for legislative action in both the province and throughout the empire to prop up the traditional mercantile system. Theoretically this would prevent the introduction of American capital and business methods into the local economy. Illegal practices such as employment and smuggling among the fishing laborers, however, greatly limited the effect of this theory.

Virtually every participant in the North Atlantic fisheries utilized the truck system at some point in time. When European colonials first established a resident fishery in North America, they had neither the capital nor the labor force to compete with migratory European-based firms.[4] In addition, the immediate presence of a large amount of cheap land made it difficult for early American fish merchants to retain their labor force. In order to prevent a larger migration to the agrarian sector, these merchants needed to develop a system whereby their laborers would be dependent upon them. Likewise, due to the shortage of currency available in the colonies, they needed to reduce the risks of fishing voyages and limit their capital investment in that industry. The truck system answered these needs.[5]

Instead of extending capital to be invested in the industry, merchants extended credit to fishermen. The fishermen could use this credit to purchase the food and tools necessary for the voyage. This not only limited the amount of raw capital that each merchant needed for the industry, but it also answered the need of the fishermen who had no capital to invest. Merchants also guaranteed the fishermen a continual supply of necessities throughout the winter. In return, the fishermen would supply the merchant with their catch, which the merchant posted to their credit at a deflated rate. As a result, this system assured the merchant that he would receive something for his investment, and need not fear that a competitor would take the product. This practice reduced the financial risk of the voyage. Finally, the system tied the fishermen to the merchant by debt. Annual catches seldom covered the fisherman's total credit advances, leaving him indebted, and this prevented most fishermen from migrating to another industry or even another merchant.[6]

Fishing firms throughout the North Atlantic relied heavily on the debt-credit bond between the merchants and laborers well into the nineteenth century. Nova Scotia merchants used the truck system as their chief form of economic structure, but many contemporary businessmen and politicians in the province saw this system as a limitation to the potential business growth when compared with the more flexible methods used by American merchants. For example, Nova Scotia politician Gilbert Tucker argued in 1837 that this method restricted the region's development:

> Our fishing Vessels are owned by poor men, they get their out-fits on credit, at the highest possible rate—their hands are generally hired, his own spirits are dulled from the knowledge of the disadvantageous circumstances under which he has to labour, his hands have the same feelings, in some measure, with the additional one, of the uncertainty of being paid, thence their want of energy and the unprofitableness of our fishing.[7]

This lack of economic incentive, some argued, limited the ability of Nova Scotia's fishing industry to compete successfully in the global market.

The reality of the world market made demands on the mercantile system that it could no longer meet. While these fishing operations in the provinces bordering the Atlantic held tightly to their traditional ways of running the fishing industry, British authorities began to deconstruct the protective system of mercantilism that Nova Scotia fish merchants depended upon so heavily. Following the Napoleonic wars, the British Empire further adopted the philosophy of free trade and gradually opened several of its ports in the West Indies, by far the largest market for Nova Scotia's fish merchants, to American-based fishing firms.[8] Meanwhile, rapid population growth and the beginnings of an industrial age in the United States created an even larger market for fish and fish products for its growing urban work force. Thus, by the 1830s the international advantages in fish marketing favored the firms in the United States over those in British North America.[9]

Nova Scotia's merchants resisted the opening of the free ports from every possible angle. Their enduring belief in the benefits of traditional mercantilism became the basis of their objections, and in letters to King George IV, the Assembly of Nova Scotia attacked plantation farmers in the West Indies for threatening the destruction of the mercantile system. One letter in 1822 stated:

> Some of your Majesty's Subjects are united with Foreigners, in endeavouring to change a system which Your Majesty's Government has pursued for some years, with so much advantage to all Your People who are interested in the permanent welfare and prosperity of Your Dominions in North America, and the West Indies.[10]

According to these politicians, transatlantic trade of North Atlantic fish, British North American agriculture, West Indies sugar, and British manufactured goods benefited all British subjects throughout the world. One petition succinctly noted: "Your Majesty's Loyal Subjects in North America have no desire to advance their local interests at the expense of those of the Empire in general, but humbly conceiving that in the present case, the general interest is identified with theirs."[11]

These Nova Scotia politicians suggested that the mercantile trade also benefited England's own industrial power by making the British subjects in the

Western Hemisphere "better customers every year to the British Manufacturers."[12] If the United States grew to dominate the staple trade in the West Indies, these politicians argued, they would surely also dominate the trade of goods throughout the Atlantic world, thus threatening the whole Empire. If London officials allowed the Americans into this market, "Great Britain would provide a Country, which appears destined to become her Rival, with the means of procuring Freight upon their several Voyages, and thus add to their commercial wealth and their maritime power at the expense of her own."[13]

Nova Scotia's political leaders thus endeavored to preserve their "Commercial Privileges," which they believed they possessed as subjects of the British Empire. Tables and data accompanied the letters to London to prove that the North American colonies fully supplied the British West Indies with necessary staple resources such as grain, fish, and lumber.[14] While the West Indies claimed to be deprived of these basic necessities, the Nova Scotians critiqued both the West Indies and the United States for falsely manufacturing these shortages. In another letter to King George IV, the Assembly of Nova Scotia accused those in the West Indies of improperly and inaccurately encouraging the British government to "abandon the wise regulations which excluded the People of that Country [the United States] from participating in a Trade, which it has been always the policy of the Mother Country to reserve for British Subjects."[15]

In addition to these new markets that American fish merchants found in the West Indies, the American economy itself went through an extensive period of expansion throughout the nineteenth century. Most American historians who focus on this economic development concentrate on the expansion in western agriculture. Yet this growth also profoundly affected the North Atlantic fishing industry. First, the general population increase, and the specific growth in the urban population of the northeast, created within the borders of the United States one of the world's largest markets for cheap food supplies, including fish. Second, the commercialization of American society and industry resulted in the concentration of capital and power in the Northeast, thereby giving the business leaders of the fishing industry the capital to invest in new fisheries.[16]

This investment often came in the form of new ship designs that could meet the needs of the growing industry, and many contemporary observers commented on the vast superiority of vessel design and outfitting in Massachusetts. Royal Navy officers who patrolled the fishing waters off the Atlantic Provinces were well versed in naval architecture and held no illusions about the shortfalls of local design and construction. These officers recognized the wide variety and quality of vessels in the North Atlantic. For example, Captain James Daley observed in 1853 that

> The American fishermen deserve a great deal of praise. Their vessels are of the very best description, beautifully rigged, and sail remarkably fast; well found in every particular, and carry large crews, a great many of whom are men from the provinces. The difference between the American and English vessels is very great, for all the English vessels in the Gulf of St. Lawrence the past fall, there were only four or five could in any way compete with the American. . . . I can scarcely convey to you a description of most of the English vessels; they are of the worst models, badly masted, poorly rigged, wretchedly found in sail and rigging, and about half manned.[17]

Increased profits in New England allowed for improvements in vessel design, but these improved vessels also facilitated the further growth of New England's industry; this technical improvement was thus both a cause and result of intensive capital investment and improved market orientation.[18]

The concentration of wealth in the Northeast also gave southern New England merchants greater influence in governmental affairs, and this power ensured that the federal government would continue to support the fisheries through the use of bounties on both fish and ships. Beginning in the late eighteenth century and lasting until the early 1860s, the American government set out an aggressive plan to encourage the expansion of the fishing industry. The New England fishing firms received handsome bounties based upon their total catch and the size of their vessels. In southern New England, business leaders took advantage of this bounty system and began to construct several large fishing vessels. This resulted in a concentration of capital and power in a few large corporations within the fishing industry. These corporations then invested large sums of money in the industry and expanded their domination of the fishing grounds at the expense of the more modest firms in Maine and the Atlantic colonies.[19]

On July 4, 1789, the American federal government passed the first of several acts that established a duty on the importation of salt and a bounty on the exportation of fish. Various subsequent acts adjusted the size and requirements of the awards. The Tonnage Act of 1792, for example, replaced the bounty on exported fish with a bounty based on the tonnage of the vessel used. This act gave the bulk of the bounty to the vessel owner rather than to the exporter or merchant. Under the direction of President Thomas Jefferson, whose continual interest in agrarian development often led him to oppose merchant- or manufacturing-related legislation, the federal government repealed the duty and bounty on March 7, 1803. After the outbreak of the War of 1812, the government reestablished the duty on salt in 1813. In order to assist in the paying of war debts, the salt bounty stayed on the books as a result of the Bounty Act of 1816. The last major adjustment to the legislation was the Bounty Act of 1819. This legislation required that any vessel that received government subsidies maintain a three-quarters American crew. After this legislative activity, the fishing bounties and the salt duty remained unchanged and unchallenged until the tariff debates of the late 1830s and 1840s.[20]

These debates took place in the context of a larger controversy regarding the role of the national government in economic development. Those who supported the national government's active role in maritime commerce in order to preserve national trade, and therefore national power, typically also supported the fishing bounties. Those who believed America's economic destiny was in agriculture rather than foreign trade opposed national subsidies to any maritime industry, including the fisheries. Not surprisingly, Northern politicians often fell into the former cohort, while Southern delegates favored the latter philosophy.[21]

During these tariff debates of the 1830s and 1840s, the opponents of the bounty directly linked this government grant to the duties imposed upon the importation of salt. In referring to the Act of 1792, the majority of the Committee on the Origins and Character of the Fishing Bounties and Allowances stated that

> This act is as explicit as human language can make it, in showing the allowances to the fishing vessels, as well as, the bounties to pickled fish and exported beef and pork, to be founded upon the salt duty; rising with it as a matter of course, without any recommendation to that effect. Its limited duration to two years shows that encouragement to training of mariners was not even thought of. Two years would hardly supply a nation of mariners.[22]

The majority of the committee continued to show that the fishing bounty did not relate in any way to a desire to support the fishermen exclusive of national economic needs, nor was it an attempt to raise a trained body of seamen for the naval defense of the nation.

> The repeal [of the salt duty] was considered, purely and simply, as a question of revenue; not a word being discussed by which any human being dissented from the repeal on account of keeping up the fishing bounties and allowances, or took them into consideration of the question in any shape or way whatsoever.[23]

In examining the earlier attempts to repeal the duty on salt, and with it the bounties awarded to the fishing industry, the committee noted that the failure to repeal the legislation was not based on the encouragement of the fisheries but instead upon the fiscal needs of the government.

In addition, the opponents argued that the fish bounties and vessel allowances went against the nature of American industry, and that this legislation interfered with the self-promotion and individual will that allowed an industry to succeed in a capitalistic society that endorsed free enterprise. The legislation, opponents argued, resulted in "a depression of energy, induced by the protection of government."

> If our fishermen desire to obtain control of the home market, they must do it by increased skill and by making a better use of all natural advantages than they ever made before—not by appeals to Congress for legislative aid. . . . Let the bounty be repealed; let these Fishermen be compelled to depend upon their own energies like other people, and their pursuit will be successful enough, and probably realize some of the advantages which they claim for it. But experience of every nation shows that the tendency of legislation protection is to depress individual exertion, and render unprofitable that which otherwise might be profitable. . . . The friends of free trade can no longer in good faith sustain the bounty system.[24]

The authors, however, failed to note that the government was actively aiding other economies and laborers throughout the nation via taxes, bounties, and import duties.[25]

Other opponents of the fishing bounties attacked the system based upon accusations of fraud and misuse. J. Ross Browne, from the Department of the Treasury, accused fishing-boat captains from Eastport, Maine, to New Bedford, Massachusetts, of dishonesty in their record keeping. These records provided the sole determinant for the government bonds. As Browne reported, "It is well known to officers of the customs that but little importance is attached to an oath by this class of men . . . such frauds are of frequent occurrence." The captains and crews of the vessels did not constitute the sole practitioners of this fraud. Browne continued:

> The shores are settled chiefly by fishermen; and the community is bound together by a common interest. It is alleged that during the season when halibut and other fresh fish are in demand, vessels, under cod fishery licenses, run in and dispose of their fish, or anchor about the shore and carry on mackerel fishing, without danger of discovery. The crews are usually interested in the results, or ignorant of the law; and the community, to whom such practices are familiar, consider it fair and proper to take advantage of the government, and dishonorable to become informers.[26]

Here, Brown clearly shows that the livelihood of a fishing community depended to a great extent upon the government bounty system.

Supporters of the fishing bounties argued that the bounties and allowances did not exclusively relate to the duties on salt. Instead, John Davis, the author of the committee's *Minority Report*, argued that the committee should strive to understand the spirit as well as the letter of the law. The intent of the bounty was to aid and encourage the development of the fisheries. That, Davis argued, was precisely why the government designed the fishing bounties to exceed the duties fishermen paid on salt. The extra funds assisted the fishing communities in the purchase of other necessary commodities, such as shipbuilding materials, that in 1840 still required the payment of import duties. To repeal the bounties simply based on the reduction of salt duties would thus violate the spirit of the legislation.[27]

In addition, the *Minority Report* made reference to the Act of February 18, 1792, which based the bounty on the size and condition of the vessel used. As John Davis argued, this act had no reference to the salt used and instead rewarded individuals for maintaining larger and more efficient vessels that could become useful in times of war.[28] The fishing industry's ability to provide for the defense of the nation, in both the construction and maintenance of larger vessels and in the training of the next generation of seamen, played a major role in the defense of the fishing bounties. Davis used the Bounty Act of March 3, 1819, which required a majority American crew, to argue that the authors of the legislation intended the act to support a nursery of American seamen. Therefore, he maintained, the government should not repeal the bounty based simply on a reduction of a salt duty.[29]

Those supporting the bounty also noted the democratic nature of the legislation. According to their interpretation, by providing five-eighths of the bounty to the crew and only three-eights to the owners, the bounty supported fishermen more than the owners or merchants, whom the reduction of the salt duties most directly affected. As Davis claimed,

> This, obviously is designed to allure fishermen into the business, and to make them exert themselves in it, as the share of each depends on his own success; and this interest in the voyage is one of the greatest elements of prosperity in the fisheries. If the bounty was a drawback [for the salt duties] and nothing more, its amount would be ascertained and paid to the owners like other exporters; and would be allowed only on fish exported, or, at most, upon the salt used.[30]

Relying heavily on the democratic nature of the fishing bounty system, these supporters believed that the repeal of the bounties would have dramatically ill effects upon the equality of the fisheries and the success of free enterprise throughout the entire economy.

Unlike other industries, the supporters of the fishing bounties argued, the fishery of the North Atlantic was such an unstable industry that only the wealthiest corporations in the business could sustain continual years of low profits without the government bounties. John Williams of Kittery, Maine, observed this fact when he stated,

> The bounty now paid is the sole means which many have to procure their outfits for the voyage; and that, if it be taken away, all such persons will have to abandon the business, which will thereby fall into fewer hands, and will, in fact, be monopolized by capitalists; the inevitable result of which will be a vast reduction in the quantity of fish taken, and a corresponding augmentation in the price, which will thereby drive our fish from foreign markets, when they can now barely sustain a competition with the British exporters of the same.[31]

Therefore the bounty was necessary to keep the industry democratic and open to those without large savings of capital. Massachusetts congressman John Anderson would later argue that "The profits from the business are so uncertain, both from the scarcity of the fish and low prices, that unless they had the bounty to rely upon, it would, as a business, be abandoned."[32] This sense of democratic economic development would remain an important factor in American fishery policy.

Regionalism played a key role during these debates in the Senate. When in 1839 Senator Thomas Benton of Missouri proposed a bill for the repeal of the salt duty and with it the repeal of the fishing bounties, Senator Robert Williams of Maine quickly responded. Williams was not opposed to the repeal of the salt duty; in fact he sought a complete renovation of the government's tariff policy and taxes on other articles that were, as he pointed out, "more oppressive and more unequal in its operation upon the different portions of the Union, than is the tax proposed to be repealed by this bill."[33] Senator Williams expressed concern as to the reasoning why the salt duty should be examined independent of the other duties. Williams argued that the true motivation behind Senator Benton's proposal was not the fishing duty, but the repeal of the entire fishing-bounty system.

Senator Williams drew upon Benton's obvious regional interest when he confronted the proposal, asserting that Benton had no real reason to request the repeal of the salt duty because, as a westerner, he and his constituents received domestically manufactured salt at a reduced price. Benton's bill, claimed Williams, "provides simply for the repeal of the salt duty and fishing bounty, for the benefit of his constituents, and thereby taking from mine the only item in [the] whole legislation which goes, in any degree, to compensate my section of the country for the most grievous and unequal operation of the present tariff."[34]

As a Maine representative, Williams claimed to represent the laborers and working poor of the East who did not have the luxuries of the wealthy West. Williams claimed that Benton's voting record, which favored huge land grants to railroads, clearly placed Benton in favor of the rich corporate elite of American society. Williams stated:

> But to test this objection by the acts of the honorable Senator, I would ask how he can contend that the bounty to fishing vessels is unconstitutional when he advocates and votes for bills to grant lands in aid of the

> construction of canals and railroads; to authorize States and incorporated companies to import railroad iron free of duty; and to authorize a drawback of duties upon such iron upon proof that it has been used in construction of railroads? Is there any principle upon which such gratuities to State and incorporated companies can be defended, which would not apply with more justice and propriety to the bounty to poor fishermen? The Senator surely will not be willing to be resented as advocating and voting gratuities to incorporated companies which he would deny poor individuals.[35]

Thus Williams presented himself as the representative of the humble fishermen. He characterized them as "a hardy, industrious, and adventurous population" who "resort to the grand banks which never stop payment" in order to provide food for the masses.[36] The bounty had its most dramatic effect on these smaller fishing operations by providing them with the badly needed initial capital. The bounty could thus be used to revamp an old vessel, or as a down payment for a new and larger schooner.

This defense of the honest and humble fishermen of New England would become increasingly difficult to maintain as more and more fisheries began to industrialize. By the late 1850s, large-scale merchants in Boston controlled the vast majority of the capital invested in the North Atlantic. The Reciprocity Treaty of 1854 with Great Britain initiated a system of free trade with the British colonies of North America, including the lucrative fishing operations in the Atlantic colonies. This free-trade system ensured that foreign exports of fish were routed to Boston's extensive markets, which now began to outpace those of the many smaller ports in the North Atlantic. This perpetuated Massachusetts's control of the industry.

The larger firms of Boston had no real need for the bounty, because they had the capital to sustain temporary depressions in production. As a result, the larger firms had no vested interest in the bounties and often actively sought to eliminate them because any money from the bounty would be divided among the crew and not be placed directly into the merchants' accounts. Thus, as historian Wayne O'Leary stated, "When the large fishing capitalist turned against the bounty, the last hope of retaining it disappeared. The hard reality for small entrepreneurs and working fishermen was that by 1866 the fish merchants of Boston, Gloucester, Newburyport, and even Portland had a vested interest in not opposing repeal."[37] These fishing capitalists successfully captured the New England fishing industry by preventing the development of any small-business competition. Any government bounty would aid these competitors, and thus by the end of the 1850s, the larger firms stopped supporting the bounty system. By 1857 the salt duty and the fishing bounty came under reevaluation and were finally repealed in 1866 because the bounty no longer received support from the politicians of southern New England.

Fishing bounties and vessel tonnage allowances also played a role in the governmental debates in Nova Scotia. Although the first general assembly met in Nova Scotia in 1758, it did not wield much influence until it increased its representative body.[38] The first massive migration from New England to Nova Scotia began with the Royal Proclamation of 1763, which closed the regions west of the Appalachian Mountain range to European settlement.[39] Historians estimate that during the course of the American Revolution, up to thirty thousand Loyalists arrived in Nova Scotia, most of whom had left New England and brought with them both their interests in the fishery economy and representative politics.[40] Atlantic fish merchants profited greatly during the years of the French Revolution and the Napoleonic Wars. Debate concerning government aid for the fishing industry began when those wartime profits evaporated with a revived trade depression in 1817.[41] The drop in trade with the British West Indies in 1820 only heightened this demand for governmental support. The Committee on the Subject of the Fisheries of the Province initiated the debate on a bounty system to encourage the industry. In 1821 Representative Graffe declared that the committee was "satisfied that this valuable branch of Trade requires some additional encouragement and protection."[42] Nonetheless, they did not create protection for the fishermen, but instead for the merchants who, due to the heavy duties on fishermen's salt, lost their clients to the ports of New Brunswick, where no such duty existed. As Graffe observed, the relatively inexpensive provisions in New Brunswick "induced a large portion of the Fishermen in the Western part of this Province [Nova Scotia] to go to the Port of St. John."[43]

In this way, similar to the American debate, the request for a fishing bounty had a direct relationship with the price of salt. In the opinion of the committee, such a bounty "would induce the Fishermen, in the Western part of the Province to procure their supplies from hence, and in return bring their Fish to the Port of Halifax, instead of shipping them to New Brunswick."[44] This would not lessen the burdens on the fishermen, or otherwise improve their status, because they could still purchase salt at the same price offered by New Brunswick merchants.

It would only redirect their purchases and other economic ties from merchants in New Brunswick to their counterparts in Nova Scotia. This policy, therefore, should not be interpreted as an aid to the fishermen, but instead as a supportive measure for Nova Scotia's merchants.

In the wake of this report, a bill concerning the adoption of a bounty system narrowly passed on March 17, 1823. It issued a bounty based on vessel tonnage and the exportation of marketable codfish from the province, with the stipulation that vessels of British ownership carry the cargo.[45] Subsequently, on March 10, 1824, Brenton Haliburton and the committee recommended a bounty based on the use of salt. It passed by a slim majority of two votes, fifteen to thirteen.[46] Within a year, the bounty on vessel tonnage lapsed and New Brunswick authorized a similar bounty, which, according to Lawrence Harthshore, the new chair of the Committee on the Subject of the Fisheries of the Province of Nova Scotia, had the "effect of inducing very many of our fishermen to resort thither for their supplies, and will withdraw from this country a large portion of its staple export."[47] Although the House of Assembly debated the recommendation presented by Harthshore for some time, it failed on March 8, 1826. This left the fishing industry in Nova Scotia with no governmental support.[48]

A year later, a similar vessel allowance passed with the accompaniment of a one shilling bounty for every quintal of exported codfish to European and South American markets, provided they cleared from ports in Nova Scotia and aboard Nova Scotia vessels.[49] The new committee chair, James Boyle Uniacke, who had recently joined the Reform movement, believed that although the design to increase the exportation of "Merchantable Fish" had "been fully realized," the restrictions as to the means and whereabouts of this exportation would, in the end, ruin any progress made in that industry.[50] The restrictions stood, therefore, in "direct opposition to the Spirit of the Law." The committee recommended opening this trade to ports not mentioned in the previous bill, and allowing the trade to be carried out by "Vessels of all Nations and Markets, because, so far from giving to the Fisherman an extensive market and an increased demand for the Article of his Industry, it (the current bill) confines him within a narrow limit, to the enterprise of a few British Vessels engaged in a specific Trade."[51] This action appears to be the first legislative attempt made by the House of Assembly to extend the advantages of the bounty system directly to the fishermen rather than the merchants. Since few fishermen actually owned vessels large enough for foreign trade, the act allowed the fishermen to export their products in vessels of other nations or provinces. However, the Executive Council opposed the bill. As a result, the restrictions remained in place. The assembly made other attempts to encourage the fisheries through the use of bounties, but few survived the heated debates between the assembly and the council. Those that did were rather limited in nature.[52]

The increasing size and efficiency of the American fishing fleet, supported by a government bounty and protected domestic markets, expanded their geography of operations and entered Nova Scotia waters by the 1830s. This fleet provided local Nova Scotia fishermen with yet another potential buyer, seller, and/or employer. The presence of this American fleet along the shoreline of Nova Scotia further encouraged some merchants in Nova Scotia to seek governmental aid for the fisheries. According to one petition, a bounty was necessary to compete with the Americans, who caught mackerel on the coast of the province and in the Gulf of St. Lawrence. Without governmental aid, the merchants faced a "competition that the fisheries of Nova Scotia [could not] withstand."[53] Armed with such petitions, the committee requested a renewal of the bounties. It reasoned that

> a great benefit would result by inducing many of those engaged in the Coasting Trade to return to the occupation they have been trained to, and therefore recommend that a tonnage duty should be allowed on all vessels employed in the catching of Mackerel in deep water, thus offering competition to the Americans in the fishery, and producing a force to aid in checking their lawless practices on our shores.[54]

Competition with the growing American fleet and opposition to the illegal fishing in domestic waters by that American fleet would come to dominate the debate concerning fishery management in Nova Scotia and other British Atlantic colonies.

During these early debates concerning the government's support of the fishing industry, those in favor of the bounties struggled to have them introduced for extended periods of time. These bounties became part of a larger constitutional debate between the elected assembly and the appointed Executive Council. The fishing bounties were only one of the many contested issues in the changing political system in Nova Scotia during the first decades of the nineteenth century. On the road to responsible government, the fishery's issues became enmeshed with a host of seemingly unrelated issues that included education, religion, and road construction. At the heart of these and other debates was the issue of the allotment of funds, and the question of which branch of government would control the treasury: the appointed councilmen or the elected assembly.[55]

Although both the Executive Council and the elected Assembly of Nova Scotia agreed that the fishing industry needed assistance, they disagreed as to the terms and condition of such aid. The assembly called for a bounty to be allotted directly to the fishermen upon landing, based on the size of their catch. The council, fearful of a poor quality of fish and the need to strengthen international competition, argued for a bounty to be awarded to merchants for the export of quality fish to profitable West Indian, European, and Mediterranean markets.[56] Neither the council nor the assembly agreed to pass the other's proposals for fear of losing influence in the administration of the government and control of the financing. As a result, no successful bounty would be awarded. The fishing bounties of Nova Scotia became a casualty of political debate until the introduction of responsible government in 1848, when the assembly gained control of the government. Once the assembly gained control over the executive with "responsible government," they could establish a bounty system.[57]

Historians typically considered the shift to responsible government in British North America and the dismantling of the British imperial mercantile system as two interrelated issues that played a central role in the transition in colonial government that ultimately led to the formation of responsible government.[58] In his review of the historiography on the British Commonwealth, Phillip Buckner stated that historians have equated free trade with responsible government, largely by illustrating the link between mercantilism and the executive branches of the colonial government. Buckner argued that once the assemblies of British North America gained control, they forced the creation of freer economic relationships outside of the traditional mercantile system—"thus in rapid succession both the economic and political foundations of the old Imperial system were replaced by structures more appropriate for the future commonwealth of self-serving nations."[59] The policymakers in Nova Scotia at once called for more localized control of their internal affairs and the depression of imperial influence in the local government, while they lobbied for the continuation of an aggressively protective commercial system between the British colonies.

Neither the road to responsible government nor the development of free trade within the British imperial system suggested a desire in London to reduce its global influence. Responsible government was a device used to retain imperial control over the colonies, a political maneuver designed to secure the loyalty of the colonial elites in the continuation of imperial rule. Equally so, colonial officials accepted limited restrictions on their ability to govern because of other advantages to remaining within the Empire. In short, responsible government in Nova Scotia meant continued loyalty to the Empire with a restructuring of administrative responsibilities, thus allowing some local control while the Empire retained its authority in external relationships.[60]

The traditional representative government that characterized the first British Empire began to lose favor as the colonies in British North America modernized into sophisticated nation-states.[61] The duties of the assembly government quickly expanded beyond simple administrative tasks to include education, church patronage, road construction, and law.[62] With this modernization, the elected assembly believed they should have more initiative power in legislation.[63] The executive councils exercised patronage in British North America to control most of the government's activity. Governmental appointments in Nova Scotia were limited, and therefore the council had much less power than its counterparts in the Canadas, where government patronage was much more extensive. Nevertheless, in Nova Scotia, reformers such as Joseph Howe questioned the imperial appointment of council members who had no responsibility to the popularly elected assembly.[64]

The colonial governors in Nova Scotia, as in all the British colonies in North America, initially opposed the growing influence of a reform-minded group within the political system. In 1833 the British parliament offered to reduce its financial dealings within Nova Scotia by giving up some of its tax revenues and bureaucratic positions provided that the assembly would compensate those who would lose their posts. The new governor, Sir Colin Campbell, however, reported to the colonial office that only a small, self-interested group backed the reform movements in Nova Scotia, and thus did not represent the majority of the population. This line of defense collapsed in the election of 1836, when the reformers took a majority of seats in the assembly and consolidated their force into a working Reform Party. In 1838 Lord Glenelg, the colonial secretary, ordered Sir Colin Campbell to appoint reformers to his council. Although he continued to resist, the reformers again won the election of 1840 and Campbell conceded.[65]

Once the reformers gained complete control of the Executive Council after the election of 1848, they were able to transform it into a cabinet of ministers that were collectively responsible to the assembly for all the major decisions of the government and were composed entirely of members of the party that controlled the assembly. Thus, after 1848 the locally elected body of politicians, now in control of both the executive and the legislative councils, would control local government activities, such as the fish bounty, with little interference from London.[66] This transition in local government in no way reflected the changing attitude in regard to the position that Nova Scotia should play within imperial commerce, including the fish trade. Since reformers in Nova Scotia did not openly rebel against the Crown, as in the Canadas, the loyalty of the reformers never became an issue. Throughout the debates concerning responsible government, the colonial elites in Nova Scotia maintained their allegiance to mercantilism.[67]

For politicians in Nova Scotia, the move to responsible government included the maintenance of the traditional British mercantile system. With its new influence, the Assembly of Nova Scotia moved to protect domestic waters and domestic markets from the illegal encroachment of foreign fishing vessels, principally those of the United States. Throughout the 1830s and 1840s, fish merchants sent petitions to the House of Assembly in an attempt to make their political leaders aware that "great losses are annually sustained, in consequence of lawless persons resorting to the fishing stations."[68] The petitions repeatedly identified these "lawless persons" as "American Fishermen."[69] The petitioners called for an immediate response to the "unjustifiable interference of foreigners on Harbours belonging to His Majesty."[70] This practice, they argued, directly violated the Convention of 1818, which outlawed American fishing vessels from engaging in commercial practices and from fishing "within three marine miles of any of the Coasts, Bays, Creeks, or Harbours, of His Britannic Majesty's Dominions in America."[71]

Some local inhabitants and investigators soon recognized that American encroachment depressed the annual catch of the domestic fishing industry. They believed that this "deprive[d] the Inhabitants of a legitimate source of wealth, and transfer to Foreign productions which exclusively belongs to this Country."[72] The petitioners sought to address the "serious loss and injuries suffered by British Subjects engaged in our Coast Fisheries by the constant encroachment of American Fishermen." Just a few years after this petition, additional petitions took a more aggressive stance and called for "sending an armed Vessel, for the protection of the Fisheries."[73]

More was at stake for these merchants than the loss of a local staple product. The merchants also feared the loss of their control over the local marketplace, for the initial side effect of American encroachment was the development of an illegal trade between the local inhabitants of the Atlantic colonies and the American fleet. Nova Scotia fishermen traded their bait catch, typically herring, to American vessels in exchange for American finished goods, thereby breaking away from the control of the Nova Scotia merchants.

In an attempt to develop a working knowledge of the situation, the Committee of the Fisheries sent out several surveys in 1837 to prominent merchants. These surveys included twenty-two questions, most of which addressed the recent arrival of American vessels to their shores. These surveys provide insight into the development of governmental policy, but not surprisingly, judging from the returns, the committee only sent these questions to merchants, and not to the fishermen who violated the laws of trade and commerce. The government only questioned the merchants who faced direct competition with the encroaching Americans and thereby shaped the result in favor of opposition to the Americans.

Most of the merchants' responses focused on the smuggling of American goods in exchange for Nova Scotia baitfish. As local merchant Gilbert Tucker stated, "The Americans catch bait, and purchase from the inhabitants on the Shores of the Province; the consequence is, they pursue their Fisheries more successfully in our waters, by getting plenty of fresh bait, without loss of time; the effects are injurious to our Fisheries, the Americans purchasing bait from the Inhabitants many times."[74] Not only did this trade aid American deep-sea fishing ventures, but it also reduced the effectiveness of Nova Scotia's fishing fleet because inshore bait fishermen reserved the best baitfish for American buyers who offered them better products and terms while the Nova Scotia merchants still relied on a company store credit as a form of payment for the bait.[75]

While the friendly relationship that developed between the American fishermen and the local inhabitants served their mutual needs, it also threatened the standing of the merchants in Nova Scotia because their economy depended upon supplying the fishermen. They feared that the Americans would usurp their service. As Thomas Small pointed out, "The American Fishermen occasionally bring quantities of Dry Goods and Groceries to the injury of our Commerce and Revenue."[76] Many Royal Navy officers recognized this practice of smuggling, and as Captain Miline stated, "This illegal trade consists of provisions brought from the United States, in a greater quantity than is sufficient for their own consumption during the season, and therewith secretly carrying on an illicit trade in every port and river."[77]

The key concern expressed by the merchants focused on the protection of the honest trader and provincial revenue. The development of an illicit trade threatened, they argued, the "mutual confidence which is always to exist between the Merchant and Fishermen of a Country, inducing the former to supply and the latter to make payments."[78] The truck system would only remain in operation as long as the labor force was tied to the managerial class through debt. The arrival of new American competitors who offered the laborers better products and better prices threatened the merchants' control over their fishermen. In their own self-interest, many merchants, like those at the D&E Starr Company, wanted to stop this practice:

> Such a traffic deprived the Revenue of a large amount of Duty, the Province of a valuable export, and the Merchant, who has supplied those people with their out-fits, of his payment—the earnings of the Fishermen are squandered in useless traffic, his credit destroyed, and his time completely lost to the Country; whereas if a stop was put to their trade in our Harbours, a much greater source of valuable export would be brought to the Capital, and the credit and means of our Fishermen would annually increase.[79]

Armed with the profits of lucrative government bounties, the American fishing schooners increased their geographic field of operations, pushing farther into prohibited waters. The arrival of the American fleet thus threatened to destroy the debt-credit relationship that ensured the merchants' control over the local labor population and provided the fishermen with another viable option to secure individual property and capital. The American option soon became a very popular one, and this illicit trade developed beyond the trade baitfish caught in Nova Scotia's harbors. It also included the agricultural products of Prince Edward Island, which became a favorite stop for many American fishermen during the season. This practice brought American finished goods into Prince Edward Island and other Atlantic provinces, thereby reducing the sales of English manufacturers. As Commander F. Egerton of HMS *Basilisk* reported, American suppliers soon became the preferred choice: "Their popularity with the people may perhaps in part arise from the introduction by them of contraband goods, or more legitimately of hard cash, of which there is a great scarcity on these shores."[80]

The inhabitants of Nova Scotia repeatedly turned this illicit trade into an even more direct relationship. In a final blow to Nova Scotia's truck system, fishermen not only abandoned their credit, but also their employment by seeking out better paying jobs aboard American vessels. Elisha Payson, a merchant and exporter from Bryer's Island, reported that the fishermen of his region habitually "go to the United States during the Fishing Season for employ because they can get more wages."[81] For many fishermen, the annual migration to the United States via a fishing schooner answered the needs for product, capital, and economic independence from local merchants.

This migration of labor and smuggling of goods was illegal in both the United States and in Nova Scotia; thus it remains largely hidden from historical documents. A lack of quantitative data, however, does not weaken the abundant evidence from government, business, and agency reports in Nova Scotia that discuss the migration of labor and the possible benefits received both from the laborers and their American employers. Merchant Paul Crowell recognized this benefit in his defense of the activities of the fishing laborers in his report of 1852:

> When arriving in the United States they generally procure good wages or should they ship on shares, their fish is taken to a market in the United States, free of duty or expense. As these vessels are generally bound to some port in Nova Scotia, those who are Nova Scotia men can take their little supplies for their families, and have them landed at their door, nearly as low as they can be purchased in the United States.[82]

By offering fishermen higher wages, payment based on cash and not credit, a chance to enter the profitable American market duty-free, and access to a bounty system that rewarded the fishermen as much as the merchant, many American schooner captains were able to recruit Nova Scotia's local labor population.

The Nova Scotia fishermen in return served an essential role in the American industry. They brought with them the skills needed to perform the task and an extensive knowledge of the coast. Many officials in the province recognized these important skills, including Lieutenant M. R. Perchell of the armed tender *Alice Roger.* Percell reported to his commander that the crews of the American schooners "are large, and part of whom are natives of this province, who are generally discharged before the vessel returns to the United States. The Americans prefer these men for their knowledge of the coast, and being better fishermen."[83] These documents record the gradual merging of the laboring population across the national boundary.

Many Nova Scotia merchants saw this employment pattern as a severe threat to their own prosperity and the development of the colony, not only because of the loss of the labor population to a foreign competitor but also because of the ability of these fishermen to conduct valuable trade negotiations for their new employers with the local communities throughout Nova Scotia. James Uniacke, a local merchant and the chair of the Committee of the Fisheries, articulated these fears in 1837:

> The Youth of the Province are daily quitting the Fishing Stations and seeking employment on board United States Vessels, conducting them to the best Fishing Grounds, carrying on trade and traffic for their new employers with the Inhabitants, and injuring their Native Country by defrauding its Revenue diminishing the operative class, and leaving the aged and infirm to burthen the Community they have forsaken and deserted.[84]

Merchants throughout Nova Scotia faced a serious challenge from the arrival of the American fleet in their immediate waters. Their reports reflect both their own business interests and their interests in preserving a local economy directly tied to traditional beliefs in British mercantilism.

Reports from Royal Navy officers offer, perhaps, a less biased opinion than those of the merchants who were directly and negatively affected by this emigration of their labor force. Although these naval officers were certainly eager to carry out their orders and defend the coast from American vessels, they were less likely to demonize the British and American fishermen who cooperated in accordance with the laws. One officer reported that the "English and American Fishermen appear to be on very good terms and no disagreements of any kind came to my knowledge."[85] Likewise, Lieutenant W. W. Bridges, of the armed tender *Bonito*, suggested that this relationship mutually benefited both the American and the British fishermen. The Americans, he argued, employed the local inhabitants as pilots because of their intimate knowledge of the waters, which was essential for a successful fishing trip. In turn, these pilots profited from their employment with the Americans in many ways. According to Lieutenant Bridges,

> The advantages offered to the Nova Scotians to embark in American Vessels, in the way of bounty, and of getting their shares of the fish into the American Market, clear of heavy duty, are very great; and numbers are constantly

> mixed up with their interests, and receiving their principal support from them. . . . At Barrington, a large number of American Vessels call on their round to eastward for the greater part of their crew.[86]

With this emigration of their skilled laborers, many Nova Scotia merchants were forced to seek out other employees. Many of these new employees came from Europe or Newfoundland; they too, however, resisted permanent attachment to the merchants. Thomas Tobin, a merchant from Halifax, claimed that once these new immigrants earned a bit of money, they would quickly leave the community for more lucrative employment in the United States. As Tobin bitterly noted, "consequently two thirds of our crews are entirely strangers every year—as respect the Natives, as soon as they become of age they procure a Whaler and commence business on their own account."[87]

While many merchants saw the flight of the fishermen as an act of treason and an abandonment of their native homes and domestic interests, others argued that the traditional truck system and the limited effects of the government bounty hindered the development of a competitive industry. These individuals attacked the government for not supporting the fishing industry.[88] Reformers, such as William Crichton, criticized the lack of support for the fishermen:

> So little encouragement is there given to our Fishermen, that they are even anxious to hire on board of American Vessels bound into the Gulf, and those being very often our best Fishermen . . . our hired Fishermen, particularly if he is a young man . . . concludes to proceed with the [American] Vessel, and a winter's residence in the States generally terminates in his becoming an American Citizen, and paves the way for others of his family and former companions to follow.[89]

As a result, the local industry leaders increasingly questioned the ideals of the truck system and sought to expand their own production through increased capitalization.

By the 1830s, merchant influence over Nova Scotia's fishing labor through the use of the truck system began to collapse. The "fish hawks" of the United States swept in and took control of Nova Scotia's most productive industry. Nova Scotia merchants grew ever more fearful of their neighbor, a "Nation of *Capitalists*," which took every commercial advantage in exploiting the waters and laborers of Nova Scotia for their own advantage.[90] The merchants continued to press the British for political and military protection. They argued that the authorities should not stand by, "for it cannot be desired that so many young men should be employed on United States Vessels, and if they are, it cannot be supposed that their loyalty will not be shaken when they constantly hear republicanism loudly applied."[91] Yet, while they petitioned for more protection of their domestic waters, they also began to invest in new forms of management within the fishery industry.

Nova Scotians, both merchants and fishermen, could not ignore the influences of American industrial capitalism. Following the 1830s, the North Atlantic fisheries entered a phase of rapid expansion. By incorporating new technologies such as the dory, trawl lines, and purse-seine nets, the North Atlantic fleets brought in larger catches every year. Between 1804 and 1819, the customs officials in Massachusetts inspected on average 20,923 barrels of mackerel per year. During the 1820s the average rose to 191,090; by the 1830s the number had jumped to an average of 224,173 barrels of mackerel every season.[92] The North Atlantic fishing grounds quickly became crowded with both large schooners and open boats. Paradoxically, this extensive exploitation brought both hostility and cooperation among the fishermen.

Thomas Small and other merchants feared the "utter destruction" of their fisheries due to the encroachment of the American fleet, and they pushed for the creation of a protective fleet to patrol the waters around the Atlantic Provinces.[93] After repeated requests from Nova Scotia, the British government agreed in 1838 to supply the province with the vessels for this protection. The government ordered the fleet "to enforce a more strict observance of the provisions of the Treaty by American Citizens; and Her Majesty's Minister in Washington has been instructed to invite the friendly co-operation of the American Government for that purpose."[94] The number of vessels operating out of Halifax fluctuated yearly as funding and expenses were adjusted, but the ships normally numbered from one to three, far too few to police the hundreds of American vessels in the area.[95]

The 1841 Report of the Committee of the Fisheries, drafted two years after the initial development of this protective fleet, stated that the limited number of vessels and insufficient funds provided for this policing force resulted in a complete failure. The extensive coast of the Atlantic colonies, which included hundreds of bays, inlets, and harbors, made it impossible for only one or two vessels to effectively patrol the area. The committee pushed for added funds and cooperation with the authorities in New Brunswick, Newfoundland, and Prince Edward Island, but with little success. These local governments were unwilling to spend the necessary funds to limit American involvement in their inshore waters—especially in New Brunswick and Prince Edward Island, where grocers made considerable money trading with American vessels.[96]

With a gradual increase in the size of the force over the course of six years, the committee members reported that they were "impressed with the beneficial results that have arisen from the protection afforded to the Fisheries by the presence of the Revenue Cutters." They eagerly recommended added support and continued funding in order to realize the full potential of their domestic fisheries.[97] Although the fleet achieved some minor victories at sea, the committee continued to push for the "employment of an additional armed vessel to aid the present Government better in protecting this valuable staple of our country."[98] The new protective fleet saw some success in checking the encroachment of the American fleet; however, between 1839 and 1851 they apprehended only twenty-seven American schooners in the waters around Nova Scotia. The authorities consigned the vessels to the Vice-Admiralty Court at Halifax, where the legal authorities auctioned off those found guilty of violating the revenue laws.[99] The Nova Scotia fishing industry experienced some mild growth throughout the years of protection. For example, in 1839 Nova Scotia exported only 19,127 barrels of pickled mackerel; in 1846, after eight years of protection, the colony shipped 82,645 barrels out of its ports.[100]

By 1852, the British authorities in London began to push the American government to sign a free-trade agreement that would include the colonies in British North America. In an attempt to persuade the American officials to set up discussions of the matter, the council in London began to enforce more rigorously the rules of the Convention of 1818 that prevented American fishermen from operating within three marine miles of the coast. The local fleet of one or two small vessels was augmented by the presence of four to five Royal Navy ships, which the Admiralty ordered north from their traditional stations in the West Indies. Such a move was a dramatic display of British power in the North Atlantic. This show of force corresponded with Britain's basic imperial philosophy during the middle of the nineteenth century, which called for the use of force only with the intention of strengthening their free markets and imperial trade links.[101]

The sudden appearance of this small fleet in 1852 flushed most of the American fishing schooners from the coastal waters of Nova Scotia. British officers boarded hundreds of vessels in search of violations, thereby leaving Nova Scotia merchants free to prosper in protected waters and home markets. As one Royal Navy officer reported to the Admiralty, "The protection this year afforded by the imperial and provincial governments has been, to a great extent, ruinous to the interest of those Americans who have visited our coast."[102]

The naval pressure proved successful for Britain's larger diplomatic aim. Two years later, the United States and Great Britain signed the Reciprocity Treaty, which introduced a modified free-trade system of staple commodities in North America. Included in this treaty was a clause allowing American fishing schooners the freedom to enter the waters that were previously outlawed by the Convention of 1818. Thus, in 1854 the short-lived protection of domestic waters came to an end in British North America, and American merchants completed their domination of the North Atlantic fishing industry. During the Reciprocity period (1854–1866), Nova Scotia's exports of fish products grew, but it is important to note that American vessels dominated this export. Nova Scotia's marine industry became a mere staple provider, and their own fishing fleet garnered only modest direct investment. While New England redoubled its fishing efforts under the less restrictive regulations of the 1854 Reciprocity Treaty, Nova Scotia provided labor and commodity. The real economic growth occurred in the New England fishing firms that provided capital investment and usurped the bulk of the region's wealth.[103]

Thus by the middle decades of the nineteenth century, the pattern of cross-border migration of fishing labor was well established. By 1880 an estimated four thousand Canadians served in New England's fishing fleet. Nearly one-fifth of all crews serving aboard Maine fishing vessels, for example, were from the Maritime Provinces. Of all the foreign fishermen serving in Maine's sea fisheries, 92 percent were Canadian-born, while 52 percent of all of Massachusetts's foreign fishermen were Canadian.[104]

The ongoing migration of the American fishing fleet into British waters, in the form of both capital and labor recruitment, challenged local authority in what the Americans increasingly perceived to be an international resource economy. As the fishermen from both Canada and the United States became increasingly integrated into one regional labor force, they united in a local interpretative struggle over resource use that often contradicted the ideas created by distant authorities in the central governments of the competing nations. When local Nova Scotia fishermen took employment aboard Yankee schooners, they gained access to better wages, profitable markets, and government support. In turn, they provided badly needed skilled labor. The Nova Scotia fishermen aboard the American schooners also facilitated trade relations with local suppliers. In their native land, the Nova Scotians could barter for agricultural goods, marine supplies, and most importantly baitfish. It was the differing economic and political patterns in the United States and British North America that facilitated the initial cross-border labor migration that would in turn later set the context of the controversial American-Canadian bait trade. The free-trade policies established by the 1854 treaty temporarily avoided the issue, but when that treaty was repealed in 1866, at the same time that Nova Scotia passed its regulatory authority to the newly created Dominion of Canada, controversy emerged, and the little herring fish of the North Atlantic became the subject of intense international dispute.

2 SEEKING LOCAL CONTROL IN AN EMERGING NATIONAL CONTEXT, 1854–1885

"Intrusion of Strangers"

THROUGHOUT THE FALL of 1876, Gloucester fishing firms prepared their fleet for the departure to the winter herring fisheries along the coast of Newfoundland. The company of John Pew and Son prepared two well-equipped schooners for the voyage. Both the *Ontario*, mastered by Peter McAuley, and the *New England*, mastered by John Dago, outfitted their crews with purse-seine nets. They left Gloucester in late November and arrived in Fortune Bay, Newfoundland, a few weeks later. They proceeded to Long Harbor, where they joined a fleet of twenty-six American schooners and approximately one hundred local vessels and boats, all of which searched the waters for the next big school of herring, which could then be immediately frozen in the cold North Atlantic and transported to New England to be used as bait in the spring cod fishery. On Sunday, January 6, the fishermen saw "bubbles" rising to the surface of the water—an unmistakable sign to the crews that the fish had arrived. McAuley and Dago worked their seines together, and in a short time they caught an estimated 2,000 barrels of herring. Upon capturing the precious baitfish, a "violent mob" of nearly two hundred local Newfoundlanders presented themselves before McAuley, Dago, and the rest of the Yankee fleet working the inshore environment to demand that they release the herring. The Newfoundlanders boarded the

Gloucester schooners and began to cut apart the purse-seine nets. One American captain threatened the locals with a revolver, but to no avail. The local fishermen far outnumbered the visiting fishermen from Gloucester, overpowered them, and cut their nets open to release the herring. The Newfoundlanders celebrated their day's victory over a burning bonfire of destroyed American seine nets.[1]

This is but one example of a growing effort among local fishermen from around the northwestern Atlantic fishing world to claim and enforce control over their local bait fisheries and the exchange of baitfish between producer and consumer. The *Ontario* and the *New England* had full rights as established by the recently signed treaty of 1871 to utilize the baitfish grounds around Newfoundland, Prince Edward Island, and the Atlantic Canadian Provinces of Canada. Local fishermen in Fortune Bay, however, were not interested in adhering to the recent changes in international law and diplomacy. Instead they sought to enforce their own constructed codes of conduct that restricted extraction to local operations while allowing foreign fishermen to purchase bait caught by the local labor force. These informal codes, which emerged in the 1830s, retained important authoritative power, despite the growing influence of national politics and international treaty negotiations that began after the Reciprocity Treaty of 1854 and the Canadian Confederation in 1867.

As has been demonstrated, the North Atlantic bait fishery was quietly developing into a major diplomatic issue between the United States and Great Britain. When Canada emerged as a semi-independent dominion in North America, the Anglo-American fishery dispute suddenly became a three-way power play between the competing nation-states. Furthermore, after 1870, the colonial governments of Prince Edward Island and Newfoundland began to insist on more independent management of their own fisheries, resisting both British control and Canadian influence in efforts to seek independent arrangements with the United States government, or more informally with American fishing interests. Local legislation competed with national law and international treaty, all while fishermen continued to navigate both the environment of the North Atlantic and the language of the multiple jurisdictional authorities. Early in this debate, Great Britain pursued a general policy of appeasement and rapprochement with the United States, much to the disappointment of Canada. The emerging Conservative Party in Canada, under the direction of Sir John A. Macdonald and his Nova Scotia ally Dr. Charles Tupper, used this lax British attitude to impress upon fish merchants in Nova Scotia, and to a lesser extent in New Brunswick and

Quebec, that the Dominion government should be entrusted with the protection of the fishery resources of Canada from American exploitation. As a result of Confederation, Canada's new diplomatic role in the Treaty of Washington negotiations, and the National Policy, Ottawa increasingly became the official policymaker for Canadian fisheries management. The international events between the repeal of the 1854 Reciprocity Treaty in 1866 and the Halifax Award of 1877 tested the power of the Canadian national agenda in fishery management in both Canada and the British Atlantic colonies of Prince Edward Island, which joined Canada in 1873, and Newfoundland, which remained a colony of Great Britain throughout the period now under discussion. This power was far from absolute, as fishermen of the North Atlantic continued to defy formal law, now set by Ottawa, in an effort to buttress their own sense of managerial stewardship through their informal codes of conduct.

The United States government sought to repeal the Reciprocity Treaty of 1854 for a number of reasons wholly outside of the fisheries. First among them was the *Alabama* Claims. The CSS *Alabama* was a Confederate commerce raider built in Liverpool, outfitted by British capital, and manned largely by English sailors. Many of the Confederate commerce raiders also found safe harbor in Canadian ports. The *Alabama* and the other Confederate commerce raiders did more damage to Anglo-American relations after 1865 than they ever did to Union shipping during the Civil War.[2] In addition to the *Alabama* Claims, American expansionism also played a role in ending the Reciprocity Treaty. More than one politician in the United States sought to simply annex Canada rather than deal with it as a commercial equal or as politically independent.[3] Finally, the emerging industrial power in the United States pushed forward new theories of protected markets, and thus opposition to North American free trade.[4] These factors, rather than anything related to the fisheries, contributed to the final repeal of the Reciprocity Treaty in 1866.

The American repeal of the Reciprocity Treaty in 1866 did generate some concern within the public sphere of the nation's news media. Oppositional papers like the *Baltimore Sun* recognized that the repeal presented both a commercial and an international problem, and warned that "the termination of the treaty revives claims and pretensions, and may reawaken irritation, and give rise to collisions endangering the tranquility, as in times past they have jeoparded [*sic*] the peace of the two countries."[5] Newspapers from around the country retold to their readers the lengthy diplomatic history of the dispute dating back to American independence, often focusing their investigation of the disputed headlands theory. It was clear that American fishermen would be prevented from operating within three miles of the coasts of British North America and Canada, but dispute remained over the right of Americans to enter bays that were wider than six miles across. Canada claimed that the three-mile line extended across the headlands of any bay regardless of the width of the bay, while the Americans argued that the three-mile line followed the coastline into open bays or other inlets, thereby allowing American operations to enter bays wider than six miles across so long as they remained outside of a three-mile line from the coast.[6]

Fishing interests in British North America were thus faced with a difficult situation. The increased trade and profits generated by the 1854 Reciprocity Treaty demonstrated to many fish merchants in the British Atlantic the value of free access to American markets. When the United States announced the repeal of those terms, these fish merchants began to mobilize efforts to reestablish North American free trade. The only bargaining chips the British North American fishing industry possessed were the values of their inshore bait fishery and their close proximity to the fishing grounds, both of which made ports in the British Atlantic important, and perhaps necessary, bases of operations for New England fishing schooners. Ottawa politicians generally believed that the best diplomatic means of achieving free access to the American market was by providing Yankee fishermen free access to Canada's inshore waters and Atlantic ports. But this exchange rested on the assumption that the American fishermen could be legally excluded from those waters, and the even bigger assumption that it was possible to physically restrict enterprising American fishing masters from illegally utilizing these waters that hugged a coast interrupted by thousands of inlets, coves, and harbors that made excellent hiding places for the American "fish hawks." British fishing interests in North America were thus left with the challenge of communicating their interpretation of restricted access as well as enforcing that interpretation in a difficult environment and diplomatic context, all while convincing their own fishermen to abandon the profitable commercial relationship they had developed under the terms of the 1854 agreement.

By the 1860s, Great Britain looked to remove itself from its responsibilities in North America. Early indications showed that the Royal Navy could not be trusted to enthusiastically enforce the exclusion of the American fishing fleet. Meanwhile, efforts emerged in the Canadas to seek unified and independent status for British colonies in North America. Canadian Confederation on July

1, 1867, resulted from diverse influences. Many historians successfully demonstrated that Canadian Confederation was a British nationalistic movement directed to counter the growing influence of American republicanism in North America. As such, these historians suggest that Confederation was as much anti-American as it was pro-Canadian.[7] The Confederation debates were clouded with innumerable political issues, but in Atlantic Canada, one of the primary issues was the fisheries industry. In particular, politicians were concerned as to how Confederation might affect the rights of American fishing schooners in domestic Canadian waters and ports and the ability to freely export Canadian fish to United States markets. Those in favor of unifying the colonies suggested that a Confederation with the Canadas would provide Atlantic interests with increased economic and diplomatic strength to counter American influences. This became especially important as it increasingly became clear that Great Britain would not enforce the restrictions as strongly as many in the Atlantic colonies would hope. For example, the pro-Confederation Halifax newspaper *Evening Reporter* suggested that "united action in the protection of our common fishing grounds; removal of disabilities between Provinces, . . . and many other considerations, furnish an argument in favour of Colonial union, in behalf of our fisheries. . . . The united supervision and surveillance of our fishing interests would be hailed as a presage of future generations."[8] The repeal of the Reciprocity Treaty and the desire to restrict American access to inshore waters in the hope of forcing the reestablishment of free markets in the United States lent itself to the support of Confederation among members of the fishing economy in Nova Scotia and the other maritime colonies.

The newly formed Dominion government, under the leadership of Sir John A. Macdonald, at first sought to limit American activities to the specific language of the treaty and prevent American operations within three miles of the coast or any headland of open bays or inlets. The *Memphis Daily Avalanch* warned its readers of the potential international crisis that would result from such exclusion, stating: "[That] the Provinces are preparing promptly and energetically to visit upon our fishermen the penalties of commercial non-intercourse, may be inferred from the action already announced by the Government of Canada."[9] Meanwhile, in its retelling of the history of the dispute, the *Baltimore Sun* reminded its readers that during the 1840s "the cloud of war gathered angrily over the waters of the Atlantic, and the hostile squadrons of America and England frowned defiance upon one another" after a similar showdown between the competing nations.[10]

Other papers reported that the Navy Department was in preparation to send a squadron to "protect American interests."[11] Free-trade and Democratic papers quickly attacked this decision. The *New Orleans Daily Picayune* criticized the initial appeal of the 1854 treaty as well as the "compound passion of anger and appetite against the Cannucks whom they wanted at the same time to punish and to absorb, to hate and to have."[12]

Although President Ulysses S. Grant and Secretary William Seward assured that there were no expectations of direct conflict, the sword-rattling in Washington was well underway. In response, Macdonald sought to win consideration for a renewal in the United States by setting up a licensing system that would allow American fishermen to continue to use the ports in Canada as bases of operation.[13] Prince Edward Island and Newfoundland both followed suit. American fishermen, however, questioned the strength of the new Dominion government to enforce the collection of the license fees, and the resolve of the British Royal Navy to assist. American fishing schooners thus continued to utilize the inshore waters and the local ports without going to the trouble of purchasing the licenses. By the end of the 1869 season, it became clear to Peter Mitchell, the minister of fisheries in the Macdonald government, that the fishery license system "has not operated satisfactorily," and as a result, "American vessels have boldly entered into our bays, creeks and harbors, and have actually crowded out the native fishermen, and fished without any regard to treaty obligations."[14]

Meanwhile, throughout 1868 the newspapers in the United States debated the value of a diplomatic solution to the problem. Most of the American press outside of New England championed the idea of a diplomatic solution and a restoration of reciprocal trade; this was particularly true of the growing faction of free traders who would come to dominate the Democratic Party by the end of the 1870s. Although it may be hard to imagine the United States and Canada, along with Great Britain, slipping into a war over access to the fishing grounds, newspapers during the 1860s clearly saw this as a legitimate concern. In May 1868 the *Troy Weekly Times* reported on the recently passed House bill authorizing the president to deploy more naval forces to the disputed region, warning that "there is no saying how soon some imprudent naval officer may involve us in a vexatious if not dangerous controversy."[15]

Despite the Canadian effort to control the situation with a liberal licensing system, it soon became clear that the American fishing schooners would conduct their business as they saw fit. This business was usually directed towards the purchase of bait and other supplies in Canadian ports. In 1869 New Brunswick fishery inspector William Venning noted this in his report to Peter Mitchell regarding the failure of the license system. He stated that in Annapolis Basin and St. Mary's Bay of Nova Scotia, numerous local fishermen deceived government inspectors concerning the presence and size of the American fleet because of the value these Americans added to the small winter herring bait fishery, which was largely conducted to the benefit of the American bait supplies for their spring and summer cod fisheries. Venning called upon the Dominion government to stop the licensing system and to arm several swift sailing and steam vessels to exclude not only American fishing efforts but also American bait-trading operations.[16]

Even anti-Confederationists, such as Timothy W. Anglin of Saint John, New Brunswick, voiced their frustration with the continued aggression of American fishing operations within Canadian waters after the repeal of the Reciprocity Treaty. In an 1870 speech, Anglin claimed that "American fishermen came where they liked, stayed as long as they liked, fished where they liked, and bore down on Canadians and drove them off their fishing grounds."[17] Although all those involved in the fishing industry could see the effects of increased American intervention, anti-confederates did not often openly complain about such action. Pro-Americanism and resistance to the confederation idea often went together. This was especially true on Prince Edward Island, where the majority of the population made considerable profits from trading with the American fishing schooners, thereby giving them closer economic ties with American fishermen than their neighbors in Nova Scotia or New Brunswick, and contributing to their resistance to Confederation.[18]

The hostility felt towards the American encroachment on inshore waters in Canada often mirrored support for Canadian Confederation, because many interested parties believed that it was the lack of diplomatic and military support from Great Britain that provided American schooners such easy access to the resources of Canada. When merchant interests in Nova Scotia called upon British authorities to reevaluate their liberal policy regarding American fishing rights in North America, the secretary of state for the colonies wrote to the lieutenant governor of Nova Scotia that

> I must distinctly inform you that on a matter so intimately connected with the international relations of this country, Her Majesty's Government will not be disposed to yield their own opinion of what it is reasonable to insist

> on, not to enforce the strict rights of Her Majesty's subject beyond what appears to them to be required by the reason and justice of the case.[19]

While merchants in Nova Scotia openly opposed the British policy of appeasement, there is some evidence to suggest that fishermen in Nova Scotia, particularly those involved in the bait fishery, sought to continue their economic ties with the American fleet despite the repeal of the 1854 Reciprocity Treaty and the restoration of the more restrictive 1818 treaty. In November 1870, Royal Navy Lieutenant Cochrane reported that "the inhabitants of the Nova Scotia coast, from St. Mary's Bay to Cape Sable, I believe prefer the Americans coming in, as they are in the habit of selling them stores, bait, and ice."[20] The bait fishery was thus shaping up to be a complicated diplomatic issue for the upcoming fishing season, one that deeply affected small-scale fishermen in Canada and the British Atlantic colonies who did not necessarily agree with their governments. The complexity was the result of the fact that not only did most American fishing operations not actually fish for the bait in inshore waters, preferring instead to purchase it from the local bait fishermen, but also this trafficking was apparently accepted by the local communities, who seemed to ignore new federal policy in favor of their more traditional informal codes of conduct that permitted trade with the Americans so long as the Americans did not fish in the local environment.

When the American fishing fleet virtually ignored the licensing system, and when American politicians showed no clear indication they would agree to new reciprocity terms, Macdonald and his minister of fisheries, Peter Mitchell, decided to enforce a very literal interpretation of the 1818 agreement by prohibiting all American fishing activities in Canadian ports, with the exception of wood, water, shelter, and repair. This policy had the immediate effect of preventing American schooners from hiring local crew members, transshipping their fish catches back to their home markets via steamers in Atlantic Canadian ports, or purchasing bait from small-scale fishermen along the coast.[21]

Mitchell directed his small navy of a half-dozen cruisers to seize American fishing schooners without incorporating the Royal Navy's longstanding practice of giving one warning before seizing any fishing vessel. In contrast to this aggressive policy being pursued by the Canadians, British governmental officials instructed the Admiralty to relax the naval protection of the North Atlantic fisheries. In 1870 Lord Granville ordered the secretary of the Admiralty to "inform their Lordships to instruct the officers of Her Majesty's ships employed in the protection of the Fisheries that they are not to seize any vessel unless it is evident and can be clearly proved that the offense of fishing has been committed, and the vessel itself captured, within three miles of land."[22] Contrary to the Canadian policy under Macdonald and Mitchell, the Royal Navy seemed content to allow American fishing vessels to hire crew, transship catches, and purchase bait and other supplies.

These instructions were forwarded to Vice-Admiral Wellesley, the commanding officer of the British Atlantic fleet. The Lords of the Admiralty also reminded Wellesley of "the extreme importance of commanding officers of the ships selected to protect the fisheries, in exercising the utmost discretion in carrying out their instructions."[23] British maritime policy of limited intervention in secure markets, like Atlantic fish, dominated both their political and their strategic policy regarding the fisheries dispute in the North Atlantic.

Mitchell strongly protested the Admiralty's decision not to capture any American vessels outside the three-mile limit. Mitchell reported to Macdonald that this policy allowed American trespassers "the anomaly of escaping from the Marine Police of Canada to the quasi-protection of Imperial authorities."[24] He concluded the report by recommending a more aggressive campaign that would include seizures of any suspected vessel regardless of its immediate location or actions. The British government, however, refrained from such an aggressive policy of exclusion; it responded with orders "to confine the action of the British and Canadian Authorities for the present to waters with respect to which no possible controversy can arise."[25] In March 1871, Earl Kimberly further undermined Ottawa's effort to claim control of the waters along Canada's coast, as well as the emerging bait trade between Canadians and British subjects and their American neighbors. In a letter to Lord Lisgar, Kimberly stated:

> I think it right, however, to add that the responsibility of determining what is the true construction of a treaty made by Her Majesty with any foreign power must remain with Her Majesty's Government, and that the degree to which this country would make itself a party to the strict enforcement of treaty rights may depend not only on the liberal construction of that treaty, but on the moderation of reasonableness with which these rights are asserted.[26]

Imperial authorities continued to push for a more passive enforcement of territorial rights in the North Atlantic and denied the Canadians any real authority to speak directly with the United States.

Earl Kimberly continued with his instructions to Lord Lisgar that so stringent an enforcement as that proposed by Dominion authorities did not reflect the general policy of the Empire towards the United States. Kimberly stated that while Canadian authorities were correct in their interpretation of the wording of the international agreement and the Dominion and imperial legislation,

> Her Majesty's Government feels bound to state that it seems to them an extreme measure, inconsistent with the general policy of the Empire, and they are disposed to concede this point to the United States Government, under such restrictions as may be necessary to prevent smuggling and to guard against any substantial invasion of the exclusive rights of fishing which may be reserved to British subjects.[27]

Clearly, imperial authority proceeded down a moderate path of appeasement. Canadian officials continued to push for more direct intervention, however, and collision with the Americans seemed inevitable.

From the beginning of the more strict enforcement suggested by the Canadian authorities, the extent to which Newfoundland and Prince Edward Island would follow suit remained doubtful. Prince Edward Island merchants gained hefty profits in their dealings with American schooners looking for ports to resupply and offload fish catches. As a result of the more stringent enforcement in and around Nova Scotia and New Brunswick, American fishermen increasingly turned to Prince Edward Island. Peter Mitchell reported in June 1871 that it was the habit of many fishing vessels from Nova Scotia, New Brunswick, and Quebec to go to Prince Edward Island, where they could sell their catch to American steamers waiting in the colony's ports. These steamers would transport the fish from those provinces as "American" fish and thus enter the American market free of duties. Mitchell argued that this business of transshipment in Prince Edward Island was chiefly the work of American merchants: "Most of the fishing trade of that Island is carried on by United States citizens, or with American capital; and large quantities of the fish marketed and shipped there, are taken by Provincial and foreign fishermen in the waters of Canada."[28]

The *Montreal Herald*, a newspaper that was generally opposed to Macdonald's government, was eager to note the impracticality of Mitchell's policy, because American fishermen were able to carry on business as usual by going to Prince Edward Island "where more liberal views prevailed." While the *Herald* criticized the island population, who had "thrown their fisheries open to foreign, which in their case means American, fishermen," the paper realized the simple practicality of the islanders' actions, noting that "the people of the Island are making money fast," and any attempt by Macdonald to force the termination of the business would not only be unsuccessful, but would alienate the island population from ever joining Canada in Confederation "if they are to be deprived of a lucrative business which they at least do not regard as a hardship."[29]

Although the government of Prince Edward Island was quick to deny the *Herald*'s accusation that they had "thrown the fisheries open" to American fishermen, they did not deny the profitable business of transshipment that emerged following the Reciprocity Treaty of 1854. Lieutenant Governor Robert Hodgson insisted that his government had always acted in unison with Canada in issues related to the fisheries: they adopted the licensing system along the same lines and with the same fees as Canada, and they repealed it, against their own better judgment, when Canada repealed it.[30]

The Executive Council of Prince Edward Island placed the blame for the trouble squarely on the Dominion's shoulders and further stated that the decision of the minister of fisheries "tends to seriously estrange a friendly but proud and sensitive nation [the United States]." The council argued that they had followed Canada's example of strict exclusion despite the fact that this excluded "from their harbors of their best customers"—customers who benefited many sectors of the island's economy from fishing to manufacturing. The council felt "bound to protest" the new policy that was "inapplicable at the present day" and directly counter to the expressed British policy of free trade and "the principles of common sense." Not only did the council doubt that the United States could be "coerced into compliance by the pressure," but further warned against "pressing an odious system upon an unwilling people"—concluding that the new policy of non-intercourse "does not obtain the moral support of the people for whose supposed benefit it is undertaken."[31]

The council, as well as the lieutenant governor, objected to the non-intercourse policy adopted by Canada because it hurt the island economy. The council turned to Edward Palmer, QC, to provide legal phraseology that countered Mitchell's claim that the 1818 treaty prohibited fishermen from transshipping their cargoes in British ports. Palmer agreed with the report previously issued by the British attorney general that stated that the 1818 treaty prohibited American fishermen from fishing, or preparing to fish, within British jurisdiction, and would thus likewise prohibit the transshipment of fish in British ports that had been caught illegally within British jurisdiction. Palmer doubted, however, that fish caught legally in the open seas could not be transshipped. He stated that the "no other purpose whatever" clause of the 1818 treaty really meant any "purposes which are really injurious or prejudicial to the trade of the Colony, or to the interests of its inhabitants." The transshipment of fish from American schooners to American steamers while in island ports was clearly not prejudicial to the interests of the island population, because this trade facilitated the growth of the colonial economy through the fees collected for such transshipments; the purchase by American fishermen of island produce, supplies, and bait; and the hiring of local islanders as crew and shore laborers. As such, the literal interpretation constructed by Canada was unjustifiable, Palmer concluded, because the transshipment of American-caught fish benefited the local island economy as much as it benefited the American fishermen.

The conflict between Prince Edward Island's economic interests and Canadian diplomatic goals often pitted local customs officers against officers of both the Canadian Marine Police and the Royal Navy. For example, on Saturday, August 14, 1870, the Gloucester fishing schooner *Clara B. Chapman* arrived in the port of Charlottetown, Prince Edward Island. There it rendezvoused with the American steamer *Alhambra* of Boston in order to transship its catch of mackerel back to the American market. Two days later, the master of the *Chapman* reported to the customs officer, William Clark, to inquire as to the legality of this transshipment. Clark was instructed by a representative of the executive to "act in the usual way, and allow the fish to be sent to Boston."[32] Thus, the crew of the *Chapman* unloaded a hundred barrels of freshly caught mackerel onto the docks and proceeded to return to the fishing grounds of the Gulf of St. Lawrence, leaving the dockworkers of Charlottetown to load the barrels of mackerel onto the *Alhambra*. On August 17, the HMS *Valorous* sailed into the harbor of Charlottetown in order to set guards aboard the recently seized fishing schooner *S.G. Marshall*. Captain Edward Hardinge of the *Valorous* was surprised to see the *Chapman* leaving port and sent his lieutenant, John Harrison, to the customs office to inquire into the matter. There, Harrison learned that Customs Officer Clark had allowed the transshipment of American fish in a British port.

Harrison pointed out that such business violated the Island Act 6 Vic. Cap. 14, to which Clark reportedly said he "had other things to do than spend his time looking through the Statutes," and that despite what those statutes might say, Clark considered stopping the transshipment to be "an unneighborly act."[33]

Apparently, many on Prince Edward Island were happy with their commercial relations with their neighborly American fishermen. Although the island certainly had a developing fishing economy based out of Rustico Bay, it did not depend upon its fisheries to the same extent as Nova Scotia and as a result did not feel the competitive pinch as much as those fishermen and fish merchants in Nova Scotia. Prince Edward Island's fishery economy was largely one of supplying other fishermen with bait and agricultural goods, not of catching and marketing fish in the global economy. This economic pattern is clearly illustrated in the case of the *E. Houghton*, seized by Captain Edward Hardinge of the HMS *Valorous*. On August 20, 1870, Lieutenant Fraser and Assistant Paymaster John Patterson boarded the *Houghton* and learned from its master, J. D. Lavie, that the vessel had last cleared from Charlottetown and had recently returned from the Magdalen Islands and Anticosti, where it traded goods—most likely agricultural produce and herring baitfish—with foreign and domestic fishermen, and received codfish and mackerel in exchange. This codfish and mackerel was then transshipped onto American steamers bound for the United States markets. Despite the fact that this trade occurred "chiefly at sea," Captain Hardinge interpreted the business as an act of illegal trade in violation of 17 Vic., Cap. 107. Hardinge argued that when Levie "transshipped goods at sea," he committed an act of smuggling—just as if he had done so within the territorial jurisdiction of Great Britain.[34] Upon visiting the collector of customs, Hardinge further learned that Master Levie had cleared from that office as a ship from Souris while his ship's paperwork indicated that the vessel was of Charlottetown, and so Hardinge added charges of falsified documentation to the seizing report.

Collector Clark, however, doubted the legality of the seizure and suggested that the *Houghton* had no dutiable goods, because "it had not been the practice for vessels to take a clearance from one port to another on this Island."[35] This incident exploded into a sharp personal conflict between Captain Hardinge and Collector Clark that mirrored the larger argument between the island government and Canadian and British authority in the North Atlantic. Clark ignored Hardinge's order to seize the vessel, stating that Hardinge "spake authoritatively" to him, even though Clark insisted he "was not accountable in my official capacity to any others than the Administrator and his Government." Captain Hardinge proclaimed that "it seems impossible to cooperate with this officer."[36] Hardinge's harassment of both American and domestic fishing vessels, his zealous enforcement of a literal interpretation of the 1818 treaty, and his exceptionally precise and even tedious interpretation of that law alienated him from most of the local island population.

Captain Hardinge was concerned about the contradiction concerning the correct port of registry for the *Houghton*, because he believed it to be but one example of the "ruse" employed by island fishermen and merchants to circumvent the prohibition of trade with American fishing operations. Throughout the fishing season of 1870, Hardinge continually boarded domestic vessels to insist that they properly identify their vessels both on their sterns and in their paperwork. On August 31, Hardinge seized the Nova Scotia schooner *Albert* of Yarmouth for possessing an inaccurate certificate of registry.[37] In several letters to his commanders, Hardinge complained that the lack of proper identification for island vessels, both in their paperwork and on the vessels themselves, indicated that the island population was conspiring en masse against him and British authority in an effort to aid American fishing schooners in their smuggling activities.[38] The more Hardinge and his fellow officers chased domestic vessels who refused to identify themselves or to heave to when called upon by authorities, the less they were able to apprehend American smugglers. In opposition to Hardinge, Collector Clark reported that he was "not aware of illicit trade being carried on by either Foreigners, or British Subjects, within the harbors, or around the shores of this Island." By September 1870, the attorney general of the island had released all the vessels seized by Hardinge because there was no "sufficient evidence" that any of the vessels had been engaged in illicit trade or smuggling.[39] The Executive Council further supported Collector Clark in his personal argument with Captain Hardinge and found Clark to have acted in full justice of his office.[40]

British authorities instructed the government of Prince Edward Island to stop the transshipments. The *Charlottetown Patriot* believed that decision was pressed forward by Canada's "arbitrary and one-sided interpretation" of the Convention of 1818. Many on the island, especially the anti-Confederation faction, strongly opposed the policy, not only because a "large business had sprung up in connection with the landing and transshipment of mackerel in Colonial ports," but also because they felt it was the Dominion authorities, not the imperial government, that pressed for this new policy, which would certainly "be as injurious to Her majesty's subjects as it will be to their American neighbors." The Dominion's "wood and water" policy, the *Patriot* proclaimed, was "heartless and extreme," and the Dominion's "monstrous" interpretation of the 1818 treaty was "contrary to the enlightenment and progress of this commercial age . . . calculated to greatly inconvenience the fishermen, and do them a cruel wrong."[41]

The *Patriot* was the island's leading anti-Confederation press and was always eager to point out the limited values of the Dominion of Canada. While the newspapers proclaimed that the island maintained its own right to protect its fisheries, it argued that "to exclude men from Colonial harbors, who come in to buy supplies, and reship mackerel in bond, is a stupid policy, which, we are happy to say, finds no favor among the people in any part of the Dominion." Even if islanders wished to follow a course of non-intercourse, the *Patriot* argued, it could do so without the interference of Canadian influence, and more efficiently than the increasingly unpopular Captain Edward Hardinge, who seemed to do no more than harass local fishermen concerning their identification "while scores of American schooners are seen every day fishing away on the North Shore."[42]

In New England, Anglophobe and Canadian annexationists like Gloucester politician Benjamin Butler quickly pounced on actions of officers like Edward Hardinge and the directions of Canada's Peter Mitchell in order to enlist both votes for the upcoming congressional election and support for American expansionism. When Canadian authorities instructed a group of Gloucester mackerel schooners to sail around Cape Breton Island to get to the Gulf of St. Lawrence fishery instead of crossing through the Strait of Canso, a passage between mainland Nova Scotia and Cape Breton Island, Butler deemed this order as an affront to American strength and pride that demanded an immediate armed response.[43] Butler received praise for his bellicose arguments from the like-minded newspaper the *Boston Herald*, which reported that the strength of the United States Navy would soon bear down on the weaker Canadian Marine Police. The *Herald* reported that "the *Nipsic* and the *Guard*, recently returned to New York from the Dardanelle Expedition, have been ordered to the fishing grounds, and the *Frolic* will sail for the Canadian waters under similar instructions."[44]

Republican papers throughout the country believed that the American government should respond to Mitchell's more aggressive stance with an equally belligerent retaliatory measure of excluding Canadian access to American markets.[45] Philadelphia's *North American* criticized Canada's enforcement strategy, claiming that many of the seizures and warnings of the Canadian Marine Police were "no better than piracy under cover of law."[46] President Grant added fuel to the fire when he delivered a message to Congress on December 5, 1870, that partially accused Canada of unjustified actions. The *Montreal Gazette* replied that "While Canada is always ready to negotiate for the settlement of all matters in dispute, it can not submit to any 'stand and deliver' arguments."[47] Grant's statements became the basis of much criticism on both sides of the border.

Oppositional papers in the United States quickly blamed Congressman Benjamin Butler for stirring up the trouble as part of a more general plan of annexation. In the American South, Butler was widely known as the heavy-handed and disastrous leader of post–Civil War occupation in New Orleans and Norfolk. The *Galveston Daily News* argued that "for a few cod-fish, Gen. Butler is desirous of plunging the nation into a war with England, and in this benign effort he has Presidential assistance."[48] Even supportive northern papers criticized the tenor of Grant's speech: the *Springfield Republican* pointed out the "belligerent paragraph respecting Canada" and referred to the whole affair as "exceedingly stupid."[49] The *Cincinnati Daily Gazette* accused Butler of using this non-intercourse policy "of making this country a nuisance to its neighbor in order to force her into annexation."[50]

After the Canadian Marine Police captured the New England schooner *White Fawn* for illegally purchasing bait and other supplies from local fishermen, more New England papers endorsed Benjamin Butler's call for a show of military strength. The *Cape Ann Advertiser* claimed that "malice and revenge were the inspiration of these acts" of the Canadian Marine Police. The paper wrote that "nothing but reciprocity of non-intercourse, shutting out the products of the Provinces from our markets, will bring them to their senses."[51] It is important to recognize that newspapers do not necessarily provide a completely accurate appraisal of public opinion, and it is certainly the case that the papers placed this particular debate within the larger context of partisan conflict in both nations, and more general ideas concerning international policies, like the tariff debates. Yet the sheer volume of reporting on the subject from papers across the continent certainly suggests that by 1870 the situation had increased in hostility as the two nations' various newspapers did rhetorical battle with one another.[52]

Considering the sheer numbers of American schooners in the area, it would be nearly impossible for Canada's police force to restrict effectively the actions of the American fishing fleet. The Saint John oppositional paper *Morning Freeman* criticized both Mitchell's and Butler's aggressive actions and words, stating that "to assume the attitude of a bully would be as absurd for Canada as it would be mean and cowardly for the United States."[53] More moderate newspapers in the United States, like the *New York Times* and Washington's *Daily Republican*, also suggested that American politics played a greater role in the growing tension in the fisheries than any action on the seas of the North Atlantic.[54]

Tempers in the New England fishing communities again flared when the Canadian Marine Police seized a Salem schooner, the *J.H. Nickerson*. The Canadian authorities seized the *Nickerson* in Cape Breton's Ingonish Bay for lying at anchor for more than twenty-four hours without reporting, even after the authorities warned its captain of possible seizure. The Canadian seizing officer, Alexander Tory, reported that the American schooner was taken for illegally purchasing supplies and bait in a Canadian port without a license. The Americans claimed that a vessel could only be detained for illegal fishing, while the Canadians argued that they had the right to enforce Dominion revenue laws that restricted the trade of fishing commodities, including bait and supplies. In addition to the *Nickerson*, the Canadians seized five other New England schooners. Canadian authorities condemned the *A.H. Wonson*, the *A.J. Franklyn*, the *Wampatuck*, the *William Patrick*, and the *Edward A. Norton* to the Vice-Admiralty Court in Halifax, Nova Scotia, for the illegal purchase of supplies or bait; none were taken for actual illegal fishing.[55]

While these captures by the Canadian authorities received much press, thanks to the vocal Benjamin Butler and others, by the end of the 1870 season Canadian officials had only seized a total of thirteen American fishing schooners. This represented a small percentage of the total New England fleet.[56] This reality had little effect on the political and editorial atmosphere in either New England or Atlantic Canada. While Benjamin Butler ranted against the Canadian Marine Police for their affront against the American people, Peter Mitchell continued to preach an unyielding approach of absolute exclusion. Mitchell issued a policy that effectively outlawed the transshipment of American-caught fish from Canadian ports to New England ports aboard railcars without legislative action. This trade had been a regular practice in New Brunswick, and many of those merchants, like those in Prince Edward Island, profited from these shipments. Mitchell's action divided the merchants in the Maritimes, and some began to cooperate with Americans in an active violation of both revenue law and international agreement. The increase in unlawful trade between some only elevated hostile feelings in both Maritime Canada and New England.[57]

It became clear to some government officials that the situation could not continue along its course if the nations wished to avoid commercial retaliation, if not open war. In the United States the political situation shifted, and the aggressive politicians such as Charles Sumner and Benjamin Butler lost favor in Washington. While this switch in the federal political atmosphere had more to do with Sumner's conflict with President Ulysses S. Grant over policy regarding Santo Domingo and the belief that the United States would benefit from open talks regarding the *Alabama* Claims, it nonetheless created a context for open dialogue concerning the fishery issue in the North Atlantic.[58]

Politicians in Canada, however, feared that the debate related to the North Atlantic fisheries and the rights concerning the profitable bait trade would be subsumed in the larger Anglo-American issues, and as a result, Great Britain would once again sacrifice Canadian concerns to appease American interests in its quest to stabilize world politics for the Empire. In an attempt to secure Canadian support for the upcoming talks in Washington, the British delegation chose to include a Canadian representative. They nominated the popular and pro-British Canadian Sir John A. Macdonald. During the talks, however, it soon became clear that Macdonald's appointment lacked real negotiating power on behalf of Canada.

With a unified Germany and Italy in Europe, Britain sought to cooperate with the United States and secure its territories in the Western Hemisphere, even if this came at the cost of Canadian interests.[59] It soon became clear that both the British and the American delegates sought to exchange the damages of the *Alabama* Claims for American fishing rights in the North Atlantic. Macdonald knew that he had only limited influence in Washington, and he therefore could only delay talks concerning the fisheries in a desperate hope that imperial interests with the United States, including the *Alabama* Claims and Atlantic trade policy, would not bury the fishery issue.[60] Macdonald's sabotage of this exchange frustrated both the Americans and the British; his popularity in Washington quickly faded. Macdonald finally convinced the British plenipotentiary, George Frederick Samuel, Earl de Gray, to only accept free fishery in exchange for free markets from the Americans. The American delegates, however, could never accept this exchange due to the popular faith in protected markets in the United States.[61]

Peter Mitchell aided Macdonald with an enormous amount of statistical data regarding the North Atlantic fish trade and the profits made by American interests in Canadian waters. These statistics overwhelmed and frustrated the American and British delegations. Macdonald's strategy proved successful, and the delegates set aside the fishery question in order to get to the more pressing issues. For the rest of the negotiations, Macdonald remained quiet and the process unfolded rather smoothly. By the end of the talks, the delegation had yet to solve the fishery question. In a final effort, the representatives agreed to open both British and American fisheries to reciprocal rights in addition to a cash award to be paid by the United States to make up the difference in value of the two nations' fishery resources. This value would be determined later in a separate arbitration. The treaty formally recognized the right of American fishing interests to operate within the inshore waters of Canada, as well as their right to openly engage in the bait trade with local Canadian and British fishermen in the Atlantic, without winning for Canada the long-sought-after free markets in the United States.[62]

Macdonald believed that the Washington negotiations exemplified the poor treatment Canada received from the British Empire in dealings with the United States. He was certain that Canada would not give a warm welcome to the treaty and it would mean the end of his political career. Private accounts of the ceremony reported that upon signing the treaty Macdonald remarked softly, "Well, here go the fisheries."[63] Newspapers throughout Maritime Canada articulated their frustration regarding the treatment of the new Dominion by the British. Many saw the Treaty of Washington as another example of how Great Britain continually sacrificed the interests of Canada to appease the United States, while others saw it as a sign of the weakness of the Canadian confederation. The treaty required ratification in both Canada and the United States, and this debate soon reflected deep regional division within the new Dominion.

The *Halifax Chronicle* reported that the Canadian parliament would surely approve the treaty because it benefited the political leadership in central Canada: "The empty treasury of Canada and the immense field for jobbery and patronage which the money will place at the disposal of the corrupt Macdonald-Cartier government will be sufficient to insure its ratification." The *Chronicle* warned its readership that Canada was about to sell off the Atlantic fishing economy to benefit central Canada trade, and thus concluded that "We will not give them [the fisheries] up without violence."[64] This threat did not amount to much in the end, however, and both Canada and the United States ratified the treaty. All that remained was the settlement of the cash award.

A variety of political issues delayed the arbitration that was to determine the financial compensation to be awarded to Canada by the American government in exchange for the free use of the Canadian inshore fisheries. Eventually the British mediated the dispute, the Canadian and American governments agreed on the judges, and the delegates met in Halifax in June 1877. While the Americans argued that the inshore waters of Canada were of no more value to Americans than the inshore waters of the United States were of value to Canadians, and therefore no cash award was necessary, it became clear to the judges that the American fishing industry won a huge victory at Washington in 1871. The Canadian team completed a more comprehensive study of the fisheries than the Americans, and this data proved to be a deciding factor in the final decision to award Canada $5.5 million in a two-to-one vote by the arbitration court.[65]

Since its formation in 1869, the Canadian Department of Marine and Fisheries, under the direction of Peter Mitchell, engaged in an exhaustive investigation of the Dominion's fishery resources. Since the actions of American fishermen so deeply affected the Canadian fishing industry, and the personal interests of Peter Mitchell, the vast majority of the department's investigations directly related to the American fleet in Atlantic Canada. Mitchell and his investigators gathered figures related to the presence of the American fleet, the value of the product it produced while in Canadian waters, the export of fish transshipped from American schooners to the American market via Canadian steam and rail facilities, and the geographic distribution of the American fishing fleet along the coasts of Maritime Canada.[66]

For example, Peter Mitchell combined the statistical data of every fishery inspector, Marine Police officer, and Royal Navy commander in the Atlantic region regarding the sighting of American vessels for every year since 1818, and concluded that between 700 and 1,200 American schooners visited Canadian waters each season. He also estimated, again through a statistical analysis of the data, that each of these vessels made an average catch of mackerel valued at $3,600 USD and a groundfish catch of $2,000 USD.[67] While such evidence was circumstantial, it remained influential in comparison to the inability of the American delegation to respond. The presentation of this data by Peter Mitchell at the Halifax Commission greatly offset the comparatively elementary studies produced by the United States Commission of Fish and Fisheries, which had only recently begun to collect data that would be relevant to the Halifax decision. Even the director of this American agency, Spencer Baird, fully admitted the vast superiority of Mitchell's evidence and presentation.[68]

The American defense at Halifax relied instead on a diplomatic history of the conflict. The delegates argued that the entire proceeding was unjustified because both the Treaty of 1783 and the Convention of 1818 provided for the American right of free and unregulated access to the inshore waters of Canada.[69] The American counter-case attacked the data provided by Mitchell by arguing that he improperly included fishing production by the American industry that was not relevant to the topic in question, such as those waters beyond the three-mile limit and open to not only Americans, but members of all nations regardless of the Treaty of Washington of 1871.[70] For the most part, the American defense lacked any substantial evaluation of each nation's fishery resources, which was the purpose of the Halifax Commission. Through their use of extensive data and their sophisticated and logical defense of their claim, the Canadian representatives defeated the Americans in the debate and won the enormous sum of $5.5 million.[71]

The American fishing population voiced outrage at the sum of the award, and many politicians seriously considered not paying the award. After much debate, key Washington politicians, including Maine's Hannibal Hamlin, determined that peace was worth the price tag, and in the end the New England fleet would see considerable profit from the use of Canada's inshore waters. In exchange for opening these waters, Canada received a substantial cash award from the United States. However, this award did little to allay the concerns of many in Maritime Canada, especially those small-scale, semi-subsistence fishermen who were directly involved with the inshore bait fisheries. This treaty attempted to superimpose a formal law allowing American access to inshore bait fisheries over the informal codes of conduct that limited such action to local fishermen.

The historical evidence already presented suggests that small-scale bait fishermen in Canada and the British colonies of the northwest Atlantic were unhappy with Macdonald and Mitchell's strict interpretation of the Convention of 1818, which prohibited the sale of baitfish to American schooners. These same fishermen, however, were no more content with the British and Canadian agreement in the Treaty of Washington or the Halifax Commission, which legalized American rights to catch their own baitfish. What the small-scale bait fishermen wanted was to prevent outsiders from catching baitfish in local waters, which would thereby give local fishermen total control of the fishery, while also legalizing their right to sell the baitfish they caught to foreign American fishermen without interference from Canadian or British authorities. When politicians and diplomats failed to secure these interests, many Canadian and British fishermen took the matter into their own hands.

It is within this larger diplomatic context that we can now return to the actions of those fishermen in Fortune Bay, Newfoundland, in 1876, which began this chapter's story. According to the Treaty of Washington, the American fleet retained the right to fish the waters in Fortune Bay, yet local legislation denied them the right to fish on Sundays. American interests claimed that the international treaty superseded municipal laws, while the Newfoundland fishermen placed their authority in local legislation, not international treaty rights. Instead, they sought not only to enforce local legislation but also to reaffirm their control of what they considered local property. The events leading up to and following what the American fishing community called the "Fortune Bay Outrage" clearly illustrate the conflict that existed between local legislation and the international treaty. They also demonstrate the ways in which different people involved in the fisheries viewed and understood the resources. While diplomats and lawyers sought to draw a precise border of authority in the North Atlantic Ocean that separated Americans from British, fishermen used the contradictions inherent in the laws to manipulate that border to achieve their own ends.

In this respect, the Newfoundlanders did not merely respond to the Americans fishing on the Sabbath Day, or even the violation of local law. Instead they reacted to a potential loss of control over the local resources and the frozen bait trade with the American schooners. The American sea captain Peter McAuley noted the economic motivations of the local fishermen when he reported his case to the American authorities:

> The only reason of said attack and demonstration by the said persons from the shore was to intimidate the American fishermen there and to prevent them from catching herring, so that the said parties on the shore of Newfoundland might sell herring to the vessels from the United States at a high price and keep the whole control of the herring fisheries in their hands, and wholly deprive the citizens of the United States from prosecuting said fisheries at Newfoundland or obtaining the herring there in any other manner than by purchase.[72]

A number of Newfoundlanders commented on how the Americans had never before tried to catch their own herring, and that this was the first season that the American fleet brought purse-seine nets. Buying the herring from the local fishermen gave those local men a source of income, and a certain degree of control over the waters and fish that they considered to be local property.

Several Americans stated that this was neither the first nor the last instance where Newfoundlanders attempted to prevent Americans from seining for fish. Charles Dagle of Gloucester described how the locals sank large rocks in the shallows and left old gillnets in the water to interfere with the setting and hauling of seines. David Malanson, also of Gloucester, observed that "It is apparent that they will obstruct any American fishery on their shores, and are not men who would know much about rights or privileges under a treaty."[73] Newfoundlanders used these methods to claim authority over the local common property.

The Fortune Bay Outrage was not a unique incident. In November 1880, John Bowie of the American schooner *Victor* confronted a large crowd of Newfoundlanders while fishing for squid to be used as bait. The Newfoundlanders destroyed Bowie's lines, leaving him no other choice but to buy his bait from local fishermen.[74] In 1880 John Dago again found himself in conflict with local Newfoundlanders while mastering the schooner *Concord* in Conception Bay. Locals forced Dago and his men to stop fishing for squid by cutting their lines. He lamented:

> Wherever I have been in Newfoundland I find the same spirit exists, and that it is impossible for any American vessel to avail herself of the privileges conferred by the Treaty of Washington; that the fishing articles of that treaty are entirely useless and valueless, and in no sense does the American fisherman receive any benefit from the treaty.[75]

These confrontations illustrate the desire of local fishermen in Newfoundland to regain control of their water, which, in their opinion, the Treaty of Washington took from them when officials in London gave away their local fishing rights to appease American interests.

Many Newfoundlanders justified their use of violence as the only means for securing their rights. One local fisherman, Mark Bolt, testified that the nearest magistrate was in St. Jacques, thirty nautical miles away. The special constable for the neighborhood, James Tharnell, arrived in Fortune Bay the following day on Monday, January 7. He stated that the destruction of the seines would not have occurred had the American captain Samuel Jacobs not threatened the locals with his revolver. He condemned the Americans for their intimidation of local fishermen, saying that "I have no power, moral or otherwise, to enforce any rules and they don't seem to care much about me."[76] To many in Fortune Bay, the only means of enforcing local law or local custom was through local crowd action. As local fisherman Richard Hendriken of Hope Cove, Long Harbor, stated in 1878, "If the natives did not see the laws carried out themselves there might as well be no laws, for there is often no one else to enforce it. It is the only way I know, and it is pretty well understood by both foreigners and natives."[77] In the end, although Great Britain continued to insist upon the right of their colonies to regulate local resources, the British government did pay $15,000 in retribution to the American fishermen for the damages from the "Fortune Bay Outrage."[78]

Yet this conflict should not be seen exclusively as a clash between nationally identified groups. The American captains at Fortune Bay testified that the very next day, the same local Newfoundland fishermen who had earlier attacked them sought to trade their own catch for American capital and goods. This desire illustrated that the local Newfoundland population was not strictly opposed to American capitalism or the presence of American fishing schooners in their inshore waters. They required only that those Americans conduct themselves according to the local law and obey certain codes of conduct. As long as these Americans continued to buy the locally caught baitfish—instead of trying to catch it themselves—and thereby support the local economy, the Newfoundland fishermen welcomed the American schooners. Once those Americans violated the informal codes of conduct, however, they would receive sharp and immediate criticism from Newfoundland's population.

The idea that territorialism was a reflection of conflicts over the control of local resources and not purely national politics is best illustrated by the continual opposition among small-scale bait fishermen towards large, highly capitalized fishing operations from away. Local fishermen voiced opposition to outside influence in the Gulf of St. Lawrence. Canadian fishery officer William Wakeham was involved in the protection of the fisheries of the Gulf of St. Lawrence for several decades following the Treaty of Washington. In the numerous reports issued by Wakeham, local fishermen often grouped large-scale fishing ventures from the United States, the Maritime Provinces, and Newfoundland together as one category. In 1879, he reported:

> It is absolutely necessary that some kind of a vessel be furnished for this service, especially on the lower part of the North Shore, where a great number of vessels from the United States, Newfoundland and the Maritime Provinces, congregate for the cod fishery. As these vessels almost all use cod seines they are constantly getting into trouble with our own hand and line fishermen, and it is utterly impossible for the local officers to carry out the law, isolated as they are, and utterly unsupported by any force to carry it out.[79]

The protection of local resources required the limitation of the rights of not only American fishermen but also of all nonresidential fishermen.

The following year, local fishermen from the Second Fisheries District of Quebec, the North Shore fisheries from the Manicouagan River to Blanc-Sablon, continued to complain to Wakeham that "outside vessels" invaded their local waters and disrupted local production. The challenge of regulating the vast fisheries of the Gulf of St. Lawrence while preserving the peace between locals and nonlocal fishermen was daunting, and Wakeham continued to request additional support from Dominion authorities to enforce the laws in the face of increasing outside pressure. He reported to his superiors in the Department of Marine and Fisheries:

> The number of outside vessels, mostly seiners from Newfoundland, the Maritime Provinces and the United States, has considerably increased this season. . . . They fish, almost all of them, within the inshore limits, and it is becoming very difficult for the inshore fishermen, who fish with the hand and line, to get along at all. There are constant complaints that the seines are swept right around, the shore boats driving them off the fishing grounds.[80]

It is clear that the local inshore fishermen complained to Wakeham not just about American seining operations. The local fishermen regarded all seiners as outsiders, regardless of the nation of origin or the port from which they sailed.

Over a decade later, William Wakeham reported the frustrations voiced by the local fishermen of the Gulf of St. Lawrence against the use of nets by outsiders within their inshore waters. Once again, the status as outsiders was not exclusively reserved to fishermen from the United States. Their fellow Canadian fishermen received opposition for their invasion of inshore waters. The sheer volume of nets used by these highly capitalized fishing vessels from Nova Scotia and New England often prevented local fishermen, who had limited capital to invest in their fisheries, from successfully competing with these outsiders. Wakeham again issued his report to his superiors stressing the local opposition to all outside fishing interests, noting that "the resident fishermen of the Magdalen Islands complain that their fishery, which is made inshore with hook and line, is being ruined by the practice of fishing with gill nets, as carried on by the vessels from Nova Scotia and the United States." Wakeham concluded that the nets completely "wall off the fish from the bay," where locals fished.[81]

Reports show that local fishing populations of the Gulf of St. Lawrence held no specific ill-feeling towards American fishermen. In fact, through extensive bait-trade networks, local fishermen often cooperated more readily with American fishermen than with other Canadian interests. In 1872 British authorities sent the Royal Navy's commander D. M. Browne to investigate complaints against the "intrusion of strangers, who had committed serious breaches of the peace, consequent on such intrusions." After an extensive investigation, Commander Browne reported to his superiors that

> the men engaged in the herring fishery about Caraquet, do not complain in the least of being molested by the crews of United States fishing vessels. The local fishermen . . . are very much opposed to, and view in the light of intruders, those belonging to the Nova Scotian and other Provincial schooners, which generally repair to the banks off Caraquet to engage in this fishery, and it is between these two parties that the disturbances complained of occur. . . . The local fishermen complain that these vessels, with such a large number of nets, entirely monopolize the banks, and, if one of the boats belonging to the vicinity attempts to anchor or set nets anywhere near them, they are immediately molested—their nets and mooring lines cut, and instances were given in which fire-arms were used.[82]

These acts, often violent in nature, pitted local fishermen against all nonlocal fishermen, regardless of nationality, when those nonlocal fishermen attempted to fish in the local environment. At times, local efforts to restrict nonlocal activities were often successful, especially when dealing with American fishermen, who became content to purchase the baitfish from the local fishing population and thus avoid the trouble that would result if they tried to catch it themselves—despite the fact that the Americans had every legal right to catch baitfish in accordance with the terms of the 1871 Washington Treaty and the 1877 Halifax Award.

These local codes of conduct that regulated both extraction and trade within the immediate communities around the North Atlantic came under direct challenge not only by the new international settlements of 1871 and 1877, but also by the emerging influence of national politics and economic policy in Canada. The successful election of Samuel Tilley, Charles Tupper, and John Macdonald in 1878 ushered in a new phase in uniting Maritime Canada to the federal government and directing the management of its fishery resources away from the local level and into the hands of the Ottawa-based Department of Marine and Fisheries. The National Policy, with its focus on tariff protection, orchestrated the redirection of resource exportation of the Maritimes away from the Atlantic world centers of London, New York, and Boston and towards the new continental Canadian leaders in Ottawa, Montreal, and Toronto. Historian Phillip Buckner concluded that "For most Maritimers . . . the 1870s was a decade of integration. Although they had entered Confederation sullen and angry at the terms of union that had been imposed upon them, by the end of the decade their anger had been considerably dissipated."[83]

This trend in centralization was certainly not unique to Canada's Atlantic fisheries, or even to the region in general. During the 1870s, numerous responsibilities were transferred from the provincial governments to the federal government and its administrative branches.[84] Some historians argue that the process of Canadian Confederation did not even begin until after the 1878 election of the Macdonald government; one claimed that that election "signaled both the completion of the institutional structure of Confederation and the beginning of the battle over its practical and philosophical workings."[85] This is not to say that the federal government wrested complete control from the provinces. But the trend of the National Policy and the political and personal philosophy of its chief architect, Macdonald, gave the federal government the right to initiate or refuse compromise with provincial authorities.

Protecting and enlarging Canada's domestic economy became the guiding principle of the new National Policy. The tariff of 1879 was strictly a protectionist policy in an era of worldwide depression in order to reduce the volume of imports, including those from the United States, that went into Canada on rather liberal terms, even after the repeal of the Reciprocity Treaty. Without equally liberal terms from the United States, Canadian protectionists such as Samuel Leonard Tilley, minister of finance, created the National Policy to force the Americans into more liberal trade talks.[86]

Macdonald often favored the idea of compromise with provincial authorities. Nevertheless, he argued that he retained the right as prime minister to ignore provincial advice in favor of national needs.[87] This political philosophy of Macdonald and his fellow conservatives, historians have often argued, was directly linked not only to their faith and pride in their British heritage and its monarchical government, but also to the political writings and philosophy of Thomas Hobbes. Both of these traditions were used to critique the emerging political philosophy of republicanism and democracy that was taking hold in the United States. In this respect, Confederation, and the emerging power of Canada's federal government, was diametrically opposed to American ideas.[88]

Most of the Canadian histories that examine the effects of the National Policy in Maritime Canada focus on the region's shift from a resource economy to a more diversified industrial and manufacturing one.[89] However, the conservative leadership of the National Policy did not forget the primary economic base of the region. Using a financial award from the 1877 Halifax Commission, Canada's federal government began to invest $150,000 annually in the Maritimes' fishing industry in the form of a bounty system. The program continued until the 1940s. The hostility felt in the United States towards the Halifax award forced the American government to repeal the fishery clauses of the Treaty of Washington in 1885, after ten years of operation. As a result, Maritimers lost their lucrative American markets and increased their ties to central Canada. This diplomatic event, in addition to the completion of a trans-Canada railway that connected Halifax with the West, elevated the value of the domestic market for the Maritimes' fishing industry.

Conservative politicians, guided by the aging Macdonald and Tupper, continued their attempts to secure free American markets through treaty negotiations. They met with little success. The failure of the proposed treaty of 1888, which will be discussed in detail in chapter 3, and the refusal of most American politicians to even discuss the issue after that, signified that Maritime fish merchants must continue to focus their marketing towards central and western Canada. As a result of these political and economic trends, the formal management of the eastern Canadian resources also moved towards the interests of Canada as a whole and away from the local population.

This political and diplomatic trend certainly did not go unchecked. Fishermen from around the Atlantic continued to construct their own system of informal codes of conduct. Fishermen in Fortune Bay, Newfoundland, opposed British policy and Canadian influence that sought to allow American fishermen to catch their own baitfish in exchange for access to American markets, or any effort in Canada or Britain to restrict the right of local fishermen to trade locally caught baitfish to American fishing schooners. Although fish merchants in Halifax may have profited handsomely from those markets in the United States, small-scale fishermen gained little to nothing from the exchange. In fact, they lost formal control over their local resources. Their only real source of stable income was to reinforce informal control of the extraction of baitfish from their local waters and thus control the market value and distribution of that essential commodity. When they lost that formal control after the signing of the 1871 treaty, they used informal action such as crowd violence to reestablish their informal codes of conduct. Despite the legality of the American action in Fortune Bay, the informal codes of conduct created within local communities trumped the international agreement between Great Britain and the United States, and American fishing vessels were constrained to purchasing the bait rather than catching it themselves. These informal codes, furthermore, did not apply exclusively to the restriction of American fishermen, but also to the restriction of all those fishermen the community deemed to be "strangers" and who therefore lacked the communal recognition of their right to access the local commons. The inshore bait fisheries were common property, but they were not completely open. Instead, the local community put significant limitations upon the rights to access those commons. Those limitations included an exclusion of all outsiders regardless of nationality, and the restriction of the modes of production to those affordable to the local fishermen. These restrictions on both the size of the work force and the tools of that work force successfully limited production and thus created a stable resource economy beneficial to the local community, which retained control. This local control came under attack not by American or other outsiders, but from their own government, which sought to continue the well-established policy reciprocating American access to free fisheries with Canadian access to free markets.

"A fisherman ought to be a free trader anyway"

On the morning of May 6, 1886, local inhabitants of Digby, Nova Scotia, watched as an unknown fishing schooner entered the gut and passed by the docks.[1] As the schooner slowly drifted past, the local fishermen of Digby turned their eyes to the stern of the vessel, eager to know her name. To their disappointment, a piece of canvas was draped over the schooner's stern, which covered the larger portion of the name and port of registry. Later, many of these locals testified they could only make out "DAV" or perhaps "DAVID J."[2] The unidentified schooner continued past Digby and moved into the inner reaches of Annapolis Basin. The customs officer at Digby sent a message to Peter A. Scott, the commodore of the Canadian Marine Police, who was then stationed across the Bay of Fundy at St. John, New Brunswick. That evening Scott arrived off Digby Gut aboard the Canadian steamer *Lansdowne*, and the following morning entered Annapolis Basin. Just before 4:00 on Friday afternoon, May 7, Scott came across the unknown vessel and discovered that it was the *David J. Adams* of Gloucester, Massachusetts.[3] During the investigation, Captain Charles Dankin of the *Lansdowne* found fresh baitfish aboard the *Adams* and concluded that the American master Alden Kinney had purchased that bait from local fishermen. Scott then ordered the seizure of the vessel for violating the Convention of 1818.[4]

This highly publicized event provides historians a rare opportunity to examine the inner workings of the complex North Atlantic bait trade between American deep-sea fishermen and local Canadian inshore bait fishermen, and what role that trade pattern played in the much larger international economy and diplomacy of the North Atlantic fisheries. Master Alden Kinney of the *David J. Adams* entered Annapolis Basin in the hope of purchasing fresh bait supplies for his offshore fishing voyage. This trade, which began during the early decades of the nineteenth century, had become a nearly universally used strategy among New England fishing operations by the end of the century. In an interview with the *Halifax Morning Herald*, Kinney succinctly stated, "We must have bait or quit fishing."[5]

These New Englanders utilized local ports in the Maritimes as bases of operations—locations in which they would purchase supplies, hire crew, and transship their catches back to their home markets. The ability of American fishing schooners to enter Canadian and British ports for these purposes was paramount to their success. Canadian officials knew this and, in 1886, sought to prohibit such actions for the larger diplomatic goal of North American free trade. Within this larger context, it becomes apparent that the seizures of 1886 had very little to do with protecting Canadian fish or fishermen from American invaders, but instead were part of Ottawa's goal to claim territorial control over the waters of Nova Scotia, New Brunswick, Prince Edward Island, and Quebec from both the United States and Great Britain, and then to use this control for a larger national agenda. The plan was to harass New England fishing operations to the point that Washington would seek to elevate the problem by signing a new reciprocity treaty. This plan had worked well enough in 1854 and 1871, but it would ultimately fail in 1886.

When the United States repealed the 1854 Reciprocity Treaty in 1866, British authorities immediately began a campaign to revitalize that trade. They succeeded in 1871 with the new Treaty of Washington. In this treaty, the United States agreed to pay for the American right to enter Canadian waters to fish. This payment was settled at $5.5 million during the Halifax Commission of 1877.[6] The amount of this award angered many in the United States who felt that the right of Canadian fish merchants to enter the American market duty-free more than made up for free fishing in Canadian waters.[7] By 1885, when the fishery clauses of the Treaty of Washington were due for renewal, American fishing interests, represented by the newly formed American Fishery Union, campaigned for their repeal.[8] These businesses were not necessarily interested in abandoning Canadian waters, but were primarily focused on preventing the free importation of Canadian fish and thus gaining complete control of the rapidly growing American market. By the 1880s, the American urban market had far outpaced the traditional fish markets in the West Indies or Europe, and therefore free trade was not a policy widely endorsed in New England's fishing communities.[9] These interests gained valuable support from a growing sector in the United States in favor of closed markets to protect American manufacturing and industrial production. On the other side of the debate stood the Boston fish buyers, represented mainly by the Boston Fish Bureau. This group desired increased importation of Canadian fish to add competition and thus increase supply and lower prices at both the wholesale and retail levels. Boston fish buyers found allies in the Democratic Party who mainly supported the theories of free trade.[10]

When the political forces opposed to free trade, mainly represented by the Republican Party, beat out their Democratic opponents, they successfully repealed the fishery clauses of the 1871 treaty. The Dominion government in Ottawa at first sought to appease the American fishing interests in the hope of winning favor that could contribute to a renewal of North American free trade. To do this, Prime Minister Sir John A. Macdonald allowed American fishermen to enter Canadian inshore waters after July 1, 1885, the end date of the fishery clauses of the 1871 treaty. This decision met with immediate disfavor in Nova Scotia, the primary waters that American fishermen utilized. Although the Nova Scotia Conservative press quietly approved of Macdonald's decision as a noble effort to avoid hostility in the North Atlantic fisheries in the midst of the fishing season, the Liberal press quickly attacked Macdonald's decision and used it as another fine example of how Confederation aided central Canada at the expense of the Atlantic Provinces.

This Liberal press accurately appraised the situation when they noted that it was not free fishing the Americans wanted, but instead the right to use Nova Scotia, and to a lesser extent Newfoundland and Prince Edward Island, as a base of operations in which they could hire crew, transship fish, and most significantly, buy supplies—including the most important item of their operation: bait. This transition in New England fishing interests in Atlantic Canada, which now almost solely relied on the use of its ports for bases and less so on the use of their waters for fishing, marks an important difference in the diplomatic context of 1886 from that of either 1854 or 1871, when New England mackerel fishermen were still interested in fishing in Canada's inshore waters.[11]

Canada's inshore fisheries were not completely abandoned by American fishermen until the twentieth century, but by the late nineteenth century, American fishermen fished less inshore and more offshore. Inshore fishing grounds became the geography of small-scale bait fishermen. When the American mackerel fishing industry, largely based out of Gloucester, switched to the purse-seine net, it slowly removed itself from Canadian inshore waters. The purse-seine net was a bulky technology that greatly increased catches. The major drawback of this technology was that it did not work well in shallow waters.[12] Therefore, the American mackerel purse-seine fishermen increasingly turned towards the open seas to catch large schools of mackerel before they went into inshore waters to spawn.[13] Thus, the American fishermen had less need to enter Canadian inshore waters for the purpose of actually fishing. Nonetheless, Canadian ports remained essential to American fishing operations, and the inshore waters remained vital as a source of bait supplies, which, by the 1880s, the Americans had happily outsourced to Canadian fishermen. So long as American fishermen retained the right to enter these waters to purchase bait and other supplies, most were content with the idea of abandoning the right to actually fish in those waters—hence the New Englanders' opposition to renewing the fishery clauses of the 1871 treaty.

Dominion and Nova Scotia authorities were well aware of this transition in the nature of the North Atlantic fishing industry. After the repeal of the 1871 fishery clauses, the Convention of 1818 again regulated the international fishery economy. Although the Convention of 1818 clearly prohibited American fishing rights in the inshore waters of British North America, the right of American fishermen to enter Canadian waters for non-fishing activities remained questionable. The specific language of the 1818 agreement prohibited American fishermen from fishing or preparing to fish within three miles of any coast, and allowed American fishermen to enter the harbors of Canada "for the purpose of Shelter and of repairing Damages therein, of purchasing Wood, and of obtaining Water, and for no other purpose whatever."[14] The Canadian authorities claimed that this was clear enough: American fishermen, by the repeal of the 1871 agreement and the subsequent renewal of the 1818 agreement, could no longer enter Canadian waters or harbors to buy bait. The American authorities claimed that the 1818 agreement was modified by the free-trade agreements between the United States and Great Britain signed during the 1830s, which would allow American fishermen to operate as traders as well, and thus enter these ports for free purchase of supplies. To add to the confusion, the Imperial Act of 59 Geo. III, Cap. 38, the legislation adopted in Great Britain to enforce the treaty of 1818, only provided prosecution and condemnation of American fishing vessels actually found fishing or preparing to fish within three miles of the coast. There was no such legal language concerning the wood, water, repair, and shelter clause of the Convention of 1818. Thus, the legal authority to prosecute American fishing vessels for buying supplies, including bait, fell to the Nova Scotia Trade Act of 1868 and the Canadian Customs Act of 1883. This added to the debate the complicated question of whether or not United States citizens were bound to colonial legislation that apparently modified an international treaty between imperial authorities and the United States. The seizure of the *David J. Adams* was immediately seen as a test case to solve the pressing "fishery question" of 1886.

In 1885 John A. Macdonald warned of a new, vigorous policy to protect not just Canadian fish, but their entire fishery economy by preventing the Americans from using Canadian ports while the United States government still prohibited the free importation of Canadian fish into the American market. In a March 1886 interview with the press, the commodore of the Canadian Marine Police, Peter A. Scott, stated that it was his intention to "seize those who are illegally fishing," and that American fishermen "can only come into our harbors for shelter, ice and provisions."[15] This language would suggest that Canadian authorities would only prevent Americans from actual fishing in inshore waters, while allowing them into ports for more than just wood, water, shelter, and repair so long as they did not stay in port for more than twenty-four hours. Yet, in an interview with the *St. John Globe* just two weeks later, Captain Scott stated: "Under this treaty United States fishermen can only enter Canadian harbors for the purpose of shelter or to procure repairs, wood or water and FOR NO OTHER PURPOSE WHATEVER."[16] It was apparent that there was no debate on the prohibition of American fishermen from fishing in inshore waters, but as late as March 1886, there was still no consensus in Canada as to the extent to which Americans would be permitted to purchase bait from local suppliers.

On March 19, the *Yarmouth Times* reported that a town meeting in West Pubnico passed a resolution stating "that if no treaty be made with the United States, the Dominion Government be asked to impose a penalty on anyone who may be found selling, or proved to have sold bait or ice having been procured inside the limit."[17] The Conservative *Digby Weekly Courier* saluted the town of West Pubnico for recognizing the core issue in the debate: bait.

> We are glad to see that our fishermen are becoming alive to the importance of preventing the American fishermen from being supplied with bait by our people. We suggested some time ago, that it would be best for our fishermen to discontinue the practice of selling them bait, because if this course is pursued for one year their vessels rigidly excluded from our grounds, an equitable treaty will seem of much more importance to them, than it does at present.[18]

Many in Atlantic Canada had already recognized that baitfish was the region's primary asset and their best hope of winning back free entrance to the lucrative American marketplace.

The Conservative press of Nova Scotia also argued that Nova Scotia, New Brunswick, and Prince Edward Island would certainly get more attention and protection from the Dominion government than they ever did or could ever expect to get from the imperial government in Great Britain. Lieutenant-Governor Richey of Nova Scotia spoke of this in a February 25 address to the provincial parliament, stating, "It is of the highest importance that the rights of our fishermen in their coast waters be effectively protected" by Dominion authorities.[19] The *Digby Weekly Courier* noted that the Dominion government had done everything it possibly could to avoid trouble, even showing great "moderation and generosity" in extending to Americans the right to access Canadian fishing grounds after July 1, 1885.[20]

The Liberal press in Nova Scotia, however, did not carry exactly the same tune. Although they agreed with a more vigorous restriction of American rights in the Maritimes, they typically presented the actions of the Dominion government as too little, too late—the *Novascotian* going so far as to call Macdonald "probably one of the coolest and most unscrupulous makers of mis-statements now engaged in the political affairs."[21] Throughout 1885, papers like the *Novascotian* routinely attacked Macdonald's government for not saying more about the fisheries question or for not more assertively seeking North American reciprocity, referring to his lack of diplomacy as "childish folly" and "foolishness."[22] To a large extent, this attack on Macdonald was just one part of a larger assault on the Conservative party that included routine attacks on the funding of the Pacific railway and the limited value Nova Scotia got from the "N.P.," the National Policy. This critique was obviously part of a much larger opposition to the concentration of Canadian economic and political power in central Canada. The maritime interests in Nova Scotia were more interested in maintaining strong economic ties with the United States, which had been its primary trading partner for some time. The attempt to build a Canadian industrial center in Ontario and Quebec carried with it protective tariffs that would have limited Nova Scotia's foreign commerce. This resulted in a general opposition to the "National Policy" of Macdonald, and thus close political ties were made between Macdonald, the National Policy, the tariffs, trade with the United States, and the fishing industry. Many Nova Scotians, including those of the *Novascotian*, feared that Macdonald would not be aggressive in his efforts to reestablish trade with the United States, because those Nova Scotians assumed that Macdonald was more concerned with Canadian industrialization—which did not need, and might even benefit from the absence of, close trade ties with the United States.

When the news came out that Macdonald had extended to Americans the right to access Canadian waters and ports after July 1, 1885, the Liberal press was quick to pounce on the opportunity to attack the prime minister. The *Novascotian* proclaimed that "A more disgraceful, undignified, unstatesmanlike, thoroughly indecent and indefensible 'agreement' was never made by anything calling itself a government than this surrender of the whole case on the part of Sir John A. Macdonald. There has never in the whole history of Canada been anything so thoroughly shameful as this."[23] This agreement, the Liberals argued, was a clear example of Macdonald's true intention of delaying reciprocity as long as possible to win support from central Canadian manufacturers, for which the Maritimes would be an easy sacrifice by forcing them to "conduct a one-sided and unprofitable trade since confederation, and notably since the adoption of the N.P. fraud," in an effort to redirect Maritime trade away from the United States and towards central Canada.[24] The Liberal press was not necessarily more pro-American than the Conservative, nor did they universally wish for more cooperation with the aggressive Yankee fishing fleet. Quite the contrary: the Liberals wanted a more aggressive exclusion of the Yankee fishermen, so as to pressure the United States government into a new free-trade agreement. The Liberals were not necessarily pro-American, just pro–American markets.

This Liberal press argued that the extension of American rights under the 1871 treaty beyond its termination date, without an equal extension of Canadian rights into American markets, had only encouraged American fishermen not to support an extension of the terms following the end of the 1885 season—and therefore, the Dominion authorities under Conservative rule lost their only real opportunity to win for the Maritimes a reciprocity agreement with the United States.[25] At times, the Liberal press in Nova Scotia went so far as to suggest that the only obstacle to free North American trade was not the American fishermen or the American Republicans, but Sir John A. Macdonald and his government in Ottawa.[26]

Like the *Digby Weekly Courier*, the *Novascotian* celebrated the actions of fishermen who sought to deny Americans access to bait supplies. On March 29, the town of Port Maitland, a "trap fishing village" near Yarmouth, passed a resolution similar to that of Pubnico, stating that "a law should be enacted by the Dominion authorities to prohibit the transportation of bait, ice, or other supplies to American vessels"—a decision the *Novascotian* proclaimed rightly demonstrated that the "principal catchers and dealers in bait . . . won't sell to the Yankees."[27] Thus, by the spring of 1886, both the Conservative and the Liberal presses in Nova Scotia called upon the Dominion government for a stronger show of force towards the New England fishing fleet. Many in Nova Scotia were most eager to see the extent to which the Dominion was prepared to shut down the bait trade, and not just American fishing efforts in inshore waters.

The Conservatives of Nova Scotia presented an equally rigid front concerning the bait trade and regularly insisted that the Dominion government should rightly enforce the rights of fishermen from interference by American schooners. Judge Savary of Digby, Nova Scotia, a former member of parliament, argued that Canada had always interpreted the purchase of bait as "preparing to fish," and to abandon that principle in 1886 would be a great blow to Atlantic Canada's fishing economy. Despite the Americans' resort to a policy that could only be understood as a "direct violation of commercial comity," Judge Savary insisted that the Dominion government should not give in to the American hostilities by allowing New England fishing schooners to enter Atlantic ports. "To give them the right to buy bait and ice here, would be giving them an immense advantage over our own fishermen," Savary concluded. "If we yield that point we practically yield everything, and our fisheries are gone."[28] Both political fronts in Canada were thus calling upon the Dominion to execute its promises of protection.

Throughout the winter of 1885–1886, the question of the bait trade slowly emerged as the primary issue in the United States press as well. In January 1886, the *Cape Ann Advertiser*, the principal press of the Gloucester fishing community, pointed out the essential character of the North Atlantic bait trade, stating:

> The inhabitants [of Canada] are only too glad to sell these small baitfish, for in many places along the coast it is the only actual money that they get, for in their own boat fishery, which is pursued along the shore, they exchange the fish they take with the merchants, receiving goods in return. Therefore, this trade with the American vessels is of the utmost importance and of great value to the British local fishermen.[29]

When the *Novascotian* stated that with "the control of the supply of bait and ice Canada holds the key of the situation," the *Boston Post* replied that the American right to purchase this bait and ice in Canadian ports was a mutually beneficial economic relationship that did not rest on any fishery treaty.[30]

Yet, the *Novascotian* argued that the purchase of bait certainly constituted a "purpose" as specifically prohibited in language of the 1818 agreement, and "If purchasing bait is a 'purpose' then it is a purpose for which the words of the treaty exclude the American fishermen from entering our harbors just as distinctly as if it had done so in express words."[31] Both the *Post* and the *Novascotian* warned that neither of their governments would tolerate any restriction of their rights, and that their fishermen "scare worth a cent." Throughout the winter months of 1885–1886, the press in Nova Scotia continued to push hard their agenda of excluding American fishing vessels from procuring bait in Canadian ports and encouraging the Dominion authorities, provided that no new treaty was signed, to immediately seize any vessel found engaged in such traffic, all for the larger goal of North American free trade. Meanwhile, the press of New England pushed hard their assertion that the Yankee fishermen had every right to seek supplies in Canadian ports so long as they did not fish in Canadian waters.[32]

Outside of New England, the United States Republican press, including the *New York Herald*, largely opposed the reintroduction of free trade and was certainly not willing to exchange free markets in the United States for free fisheries in Canada. As such, they largely supported the idea that American fishermen had no need to fish in inshore waters, but only sought the right to enter for supplies. This press interpreted the purchase of bait as an action of commerce that was equally beneficial to both American fishermen and local Canadians who relied heavily on their American customers. The restriction of such traffic, the *New York Herald* insisted, was "extreme unreasonableness and the irritating manner in which the Canadians are attempting to set up a system of non-intercourse purely vexatious to our people and at the same time injurious to their own."[33] The

American conservative press pushed hard for a show of strength from the United States government, which should insist upon their right of trade, the protection of their fishermen, and a refusal to be bullied by their northern neighbor.

When the 1886 fishing season began, American schooners kept up their policy of seeking bait supplies from local Canadian fishermen. In April 1886, two Maine fishing schooners anchored near Cape Island. The skippers of these vessels bluntly stated to a *Digby Weekly Courier* correspondent that "they intend to call for bait wherever and whenever they please."[34] In early May, the *Courier* again reported the presence of two American schooners near Digby looking for bait and offering as high as $1.50 a barrel for it.[35] Through April 1886, the Canadian Marine Police seemed content with simply chasing the Americans out with warnings, but by the beginning of May, both the Conservative and the Liberal press in Nova Scotia pressed hard for a more stringent enforcement.[36] On May 7, Peter Scott saw his first opportunity to test the new policy when he came across the *David J. Adams* with a full load of fresh baitfish. It was immediately evident that Master Alden Kinney had not caught the baitfish himself, but had purchased the supply from local fishermen, well within the three-mile line and without requesting or receiving clearance to do so from the customs officer. The *David J. Adams* thus provided an ideal test case regarding the legality of the bait trade and the right of Canadian officials to enforce the exclusion.

During the heated seasons of 1886–1887, the *David J. Adams* was the most publicly noted case in both the United States and Canada, but in addition to the *David J. Adams*, the Canadian Marine Police seized the *Joseph Story*, the *Jennie and Julia*, the *Ella M. Doughty*, and the *Highland Light*. The *Doughty* left Portland, Maine, on April 26, 1886, carrying with it a license to touch and trade from the Port of Portland.[37] The *Doughty* was headed for the Gulf of St. Lawrence when, on May 6, it was forced into Sydney, Cape Breton, on account of poor weather, where it remained until May 13. According to the master, Warren A. Doughty, the bait they had purchased in the United States had "turned soft" and was therefore unusable. Master Doughty later testified that his vessel was departing from St. Ann's Bay, Cape Breton, when local weir fishermen "made fast to me," eager to sell their baitfish to the American schooner.[38] Doughty collected the necessary funds from his crew and purchased the bait. Local fishermen, however, testified that Doughty and several of his crew came ashore looking for the baitfish and were rather secretive about it because they knew it was of questionable legality—even insisting that the local fishermen travel with them outside the harbor to actually make the exchange.[39] One of these local herring fishermen, Angus McLeod, even warned Warren Doughty to avoid St. Ann because "the Officer on shore was a very particular man about his duties and that if he was found buying bait there he would be seized immediately, he then said that his bait was getting stale and that he would have to get some somewhere or else go home without any fish."[40] Regardless of the specifics of the transaction, it is clear that Doughty had very little trouble finding local fishermen who were willing to illegally sell him baitfish. Doughty's entire voyage rested on his ability to obtain bait—without it, he was done—and so he purchased several barrels from local fishermen. On May 17, the Canadian Marine Police schooner the *Terror*, under command of Captain Thomas Quigley, seized the *Doughty* for "preparing to fish," as prohibited by the Convention of 1818, by purchasing bait within Canadian territorial limits.[41]

American fishing masters insisted that the purchase of bait was a legitimate act of commerce outside the legal bounds of fishery laws or agreements and did not constitute "preparing to fish" as prohibited by the Convention of 1818. One of the owners of the *Doughty*, Horace M. Sargent, testified that they "took out permit to touch and trade since this Fishery trouble" began. In Canada, however, the customs officers refused to recognize this license. The attorney general of Canada had begun to present his case when the trial was dropped and the vessel released on June 29, 1886. This decision to release the *Doughty* was most likely a tactic to shift the focus back to the *David J. Adams*, which appeared, by the evidence, to be a better test case to determine if buying bait constituted "preparing to fish."[42]

Of the other vessels seized in 1886, only the *Highland Light* made it to trial. On September 1, 1886, the Canadian Marine Police seized the *Highland Light*, of Wellfleet, Massachusetts, off East Point, Prince Edward Island, for fishing within the three-mile line. Unlike the *David J. Adams* or the *Ella M. Doughty*, the *Highland Light* was charged with actually fishing rather than just preparing to fish by the purchase of bait. Master J. H. Ryder admitted to the offense, and the vessel was condemned and sold at auction on December 14 without a defense. In an interesting turn of fate, the Canadian government purchased the *Highland Light* and fitted her out as a new Marine Police schooner.[43] The *Highland Light* was clearly guilty of violating the Convention of 1818 and was liable to seizure under the Imperial Act of 59 Geo. III, Cap. 38, and as such was not related to the Canadian attempt to gain territorial control over the inshore baitfish trade.

The *Joseph Story*, of Gloucester, Massachusetts, was seized off Baddeck, Nova Scotia, in April, but was released after twenty-four hours. The *Jennie and Julia* was likewise released after a short detention. The Canadian Marine Police often detained American vessels for a short period of time. United States secretary of state Thomas Bayard filed protests with the British government for the detention of the American schooners the *Mathew Keany*, *Eliza Boyton*, *City Point*, *C.P. Harrington*, *G.W. Cushing*, *N.J. Miller*, *Rattler*, *Howard Holbrook*, *A.R. Crittenden*, *Pearl Nelson*, *Everett Steel*, *Moro Castle*, *William D. Daisley*, *Marion Grimes*, and the *Flying Scud*. In addition to detainment, eight of these vessels were also fined between $375 and $1,000 for buying baitfish from local fishermen without reporting the purchase to the customhouse. Bayard also protested that the *Rattler* had been placed under guard throughout the night, and that both the *Shiloh* and *Julia Ellen* were boarded and "subjected to rude surveillance." Many in the United States viewed these detentions and fines as even more intolerable than the actual seizures.[44]

These detentions and fines represent a change in tactics by the Canadian Marine Police after the seizure of the *David J. Adams*. While the imperial legislation relating to the Convention of 1818 provided language for the seizure of vessels found fishing or preparing to fish, it made no such authorization of seizure for conducting business in Canadian ports beyond the limits of purchasing wood, obtaining water, or seeking repair or shelter. In order to enforce any actual seizure, the Canadians had to prove that purchasing bait and supplies constituted "preparing to fish" and thus violated international treaty. Due to the potential difficulty of such a legal interpretation, the Canadian Marine Police turned to provincial and dominion legislation related to customs laws, not fishing laws, for language concerning the prohibition of purchasing bait. This legislation, however, did not authorize seizure, but only detention until a fine was paid.[45] Therefore, while all the other American vessels detained in Canadian ports in 1886 were so confined for violating local customs laws, only the *David J. Adams* went through a full trial in the Vice-Admiralty Court on grounds that the local customs laws modified the international Convention of 1818, and thus the seizure was a proper punishment for purchasing bait because it constituted "preparing to fish" as prohibited by the Convention of 1818.

During the trial of the *David J. Adams* at the Vice-Admiralty Court in Halifax, Wallace Graham, the solicitor of the attorney general of Canada, argued that the Nova Scotia Trade Act of 1868 and the Canadian Customs Act of 1883, which prohibited the purchase of baitfish without authorization from the customs officers, modified the meaning of the Convention of 1818 by including the purchase of bait as an act of "preparing to fish," and thus beyond the authorized activities of foreign fishing vessels within Canadian territorial jurisdiction. Graham argued that the *Adams* was subject to seizure not just for the act of fishing but also for the act of preparation to go forth and fish. Therefore, the *Adams* had violated provincial and dominion law, as well as international treaty obligations, when it entered domestic waters for the "purpose of procuring bait that is to say herring wherewith to fish and ice for the preservation on board said vessel of bait to be used in fishing."[46] Thus, according to the Canadian authorities, American fishermen who conducted acts of trade relative to their occupation of fishing did so as fishermen, not traders, and were thus subject to the restrictions of the Convention of 1818 as enforced not only by the imperial act, but also by the Nova Scotia Trade Act of 1868 and the Canadian Customs Act of 1883.[47]

The owner of the *David J. Adams*, Jessie Lewis, asserted that the purchase of bait did not constitute "preparing to fish" under the Convention of 1818, but was instead an act of legitimate trade. Such acts of trade were permitted so long as the trading vessel held a license to trade from the United States government, which the *Adams* held. Therefore, the *David J. Adams* did not act as a fishing vessel while in Annapolis Basin, but had conducted itself as a trading vessel. As a trading vessel, the defense argued, the *David J. Adams* was not bound by fishing legislation and instead could operate under the protection of the various trade agreements made between the United States and Great Britain. The defense lawyer, Nicholas H. Meagher, opened his argument by asserting that the *Adams* "had the right to trade, so far at least as the purchase of bait wherewith to carry on her deepsea fishing outside of and beyond said limits was concerned in said Annapolis Basin or in any other Canadian port." The defense further argued that both the Convention of 1818 and the various trade treaties relative to American activities in Atlantic Canada had been signed between the United States and Great Britain and not between the United States and Canada, and therefore only the United States and Great Britain retained the authority to enforce and interpret these regulations. According to the American case, agents of Canada, such as Peter A. Scott, were not "duly authorized by law to seize said vessel and cargo." If this reasoning was found true, then neither Master Alden Kinney nor his fellow American fishermen would have any reason to concern themselves with the government of Canada or its customs legislation when conducting business with local fishermen in Canada.[48] Instead, American fishermen would be bound only by American fishing laws and imperial legislation specifically relative to the Convention of 1818.

During the trial, James B. Hill, the chief officer of the *Lansdowne*, testified that he had previously warned Master Alden Kinney of the *David J. Adams* not to procure bait from the local inhabitants, and that such action would be considered "preparing to fish." Hill testified, "We supposed that he had been there for bait, preparing to fish, and it was in consequence of information that I received that I asked him about it. I told him I supposed he knew the law, and he answered yes. He was ordered to proceed beyond the limits."[49] Hill, as well as Captain Charles Dankin and Coxswain Frederick Allen, all testified that the herring aboard the *David J. Adams* was far too fresh to have been carried from New England and must have been acquired locally. They never suggested, however, that the crew of the *David J. Adams* ever caught the fish themselves, but all involved agreed that the herring had been purchased from local fishermen.[50]

The facts of the case were not in dispute. The defense and the prosecution agreed that the *David J. Adams*, a fishing vessel from Gloucester, Massachusetts, had not fished within the prohibited area, that it had instead purchased bait from local fishermen while within three miles of the coast, and planned to go forth to the deep-sea fishing grounds that were beyond Canadian territory to use that bait in an attempt to catch codfish. The question was the application of these facts to an interpretation of the law. The court proceedings, therefore, did not address the prohibition of fishing under the Convention of 1818, but instead focused on a debate concerning the interpretation of "preparing to fish" and how that term referred to "within three marine Miles of any of the Coasts, Bays, Creeks, or Harbours of His Britannic Majesty's Dominions in America."[51] The American case argued that the term meant that one could not fish within three miles of the coast, nor could one prepare to conduct the act of fishing within three miles of the coast. This, however, did not mean that an American fishing vessel could not be within three miles of the coast, prepare to fish, and then travel outside the three-mile limit to actually conduct the act of fishing. The defense argued:

> It seems to me that when you look at the act itself and at the words "shall be found fishing," which is the governing part of the whole sentence, which is carried all through "shall be found fishing within such distance" or "shall have been found preparing to fish within such distance"—surely that only

> means a preparation within that distance to fish and not to a preparation within the distance to fish anywhere.[52]

Thus, they argued that in terms of territorial authority, Canada had a right to protect fish within their waters, but no right to limit the actions of fishermen not fishing within their waters.

The Canadian case argued that the word "preparing" related to the act of fishing, not the location of the act of fishing, and therefore it was unlawful to prepare to fish within three miles regardless of where one would eventually conduct the act of fishing. Thus, American fishermen could not cut bait, bait lines, or purchase bait or ice while within three marine miles of any coast, even if they planned to travel outside the three-mile line to actually fish.[53] The Canadians justified this interpretation because they claimed that the Convention of 1818, as well as the Nova Scotia Trade Act of 1868 and the Canadian Customs Act of 1883, intended to prevent Americans from using the shores of the Atlantic colonies and provinces as bases of operation for their North Atlantic fisheries; whether those fisheries were inshore or offshore was irrelevant. Wallace Graham claimed that the legislation adopted by the Dominion government to provide for the enforcement of the Convention of 1818 clearly supported this interpretation. "The legislature," Graham stated, "would have an object in restricting in every possible way the privileges of the American fishermen coming within our bays and harbors and to do all in their power to prevent any extension of the right that they had given them by this treaty to purchase wood and procure water."[54]

In addition to debating about the geography in which American fishermen could prepare to fish, the trial also had to define exacting what it meant to prepare to fish and how that related to the actual act of fishing. If "fishing" included both the actual act of catching a fish as well as an attempt to catch a fish, and the law prohibited both "fishing" and "preparing to fish," then the latter had to be construed as something separate than just the attempt to fish. Wallace Graham stated: "Now if we look at the context of this act we find that the expression 'preparing to fish' is used with the expression 'fishing.' I suppose it is necessary to arrive at the meaning of the word fishing, and my contention is that it is not restricted to the act of catching the fish. There is nothing about it that implies success. It includes the attempt to fish."[55] Graham continued to argue that if the verb "fishing" already included the attempt to catch a fish by the act of setting lines or net, then "preparing to fish" must mean something in addition to this, such as the purchase of bait or ice in a preparation to go forth and attempt to catch fish. Graham thus concluded that there must be a forfeiture of the vessel and cargo if there "is a preparation within the three-mile limit" to fish, as opposed to a preparation to fish within the three-mile limit.

The Americans claimed, however, that the Convention of 1818 only intended to prevent Americans from actually extracting resources from territorial waters. The American case had a great advantage in legal precedent. In 1870 Canadian authorities had seized the Gloucester vessel the *White Fawn* for purchasing baitfish from local fishermen. In 1870 Judge Hazen of the Vice-Admiralty Court concluded that the legislation related to the Convention of 1818 did not prohibit the purchase of fishing supplies but only prohibited the actual act of fishing, and so ordered the release of the *White Fawn.* In 1886 Wallace Graham responded to this challenge by arguing that the Dominion legislation of 1883 further clarified the intended meaning of the Convention of 1818 by drawing into sharper focus the justified reasons for American schooners to enter domestic water under that treaty: specifically to obtain water, wood, or to seek repair or shelter in times of distress. Although before 1883 Canada had gratefully allowed American schooners to enter domestic waters for purposes not listed in the treaty, the national government retained the right to limit, at any time they so wished, the entrance of such vessels to the four listed reasons in the treaty of 1818, and for no other purposes whatever.[56] It was thus the 1883 dominion legislation, not the imperial legislation, that clearly stipulated the purchase of bait as an act in preparation to fish. If the court found this logic accurate, then Canada would win territorial control over its inshore waters and the right to regulate all actions of American and other foreign fishermen.

Taking this angle in the argument, Graham successfully shifted the legal debate to one that focused on the protection of territorial limits in Atlantic Canada, rather than just a protection of fishing resources. Regarding the legislation, Graham stated, "It was the more important question of territorial jurisdiction and the right to prevent fishing vessels from coming into the harbor and making her preparation in respect to the harbors, bays, and coasts."[57] In the end, the court's decision on the seizure of this Gloucester fishing schooner had very little to do with fish or the protection of fishery resources and related much more to the protection of territorial sovereignty. In his response to the defense of the *David J. Adams,* Chief Justice McDonald of the Vice-Admiralty Court agreed with Graham that the focus of the decision should rest on the meaning of "preparing to fish" and its relationship to Canadian territorial sovereignty, rather than on any intention to limit the act of fishing or to protect domestic resources. In a marvelous example of judicial discourse, the court stated:

> A great deal turns on the meaning of these words "preparing to fish in British waters." I think more apt words might be used. What is the meaning of the words? Does "in the waters" refer to the fishing or the preparation? If it refers to the fishing then a preparation outside to fish within the limits would not be narrowed by the words "British waters." . . . I am only dealing with that branch of the sentence "preparing to fish." There are two elements to that, one an active element and the other a descriptive element. The preparation is the active element and "to fish" is what they are preparing to do, and now which of the two do the words "British waters" apply to?[58]

The court goes on in this manner for some time before concluding that "British waters" did in fact refer to "preparing" rather than "to fish." Making such conclusions can only come through an intense deconstruction of legal phraseology, with little to no reference to the realties of fishing in the North Atlantic or to the original diplomatic context of 1818. Instead, the court's decision was much more in tune with the contemporary political agenda of Canada.

As important as this legal discourse was to the legacy of the *David J. Adams* seizure, it did not actually unfold until well after the seizure, and the court did not finally reach a decision until the fall of 1889.[59] It was the political and diplomatic discourse in the United States and Canada that took immediate control of the situation in 1886. The American government directly responded to the seizures of 1886 primarily through diplomatic channels to London, rather than dealing either with Ottawa or the Vice-Admiralty Court in Halifax. American officials argued that the dispute was not between individual vessel owners and the government of Canada, but was a subject of international diplomacy and therefore should not be settled in a court of law. In September 1886, Edward Phelps, the U.S. minister in London, communicated to Earl Iddesleigh that

> The interpretation of a Treaty when it becomes the subject of discussion between two Governments is not, I respectfully insist, to be settled by the judicial Tribunals of either. That would be placing its construction in the hands of one of the parties to it. It can only be interpreted for such a purpose

> by the mutual consideration and agreement which were necessary to make it. Questions between individuals arising upon the terms of a Treaty may be for the Courts to which they resort to adjust. Questions between nations as to national rights secured by Treaty are of a very different character, and must be solved in another way.[60]

The logic of such an argument rested on the belief that the international treaty of 1818 was to be held above local provincial and national legislation. The determination of statute violations of international treaty was a matter of diplomacy between the contracting parties, not a matter of jurisprudence within a subdelegation of one of the original contracting parties.

Within this diplomatic context, the United States government denied that Canada had any international legal right to regulate American vessels in the Atlantic fisheries, or to enforce or interpret any treaty between Washington and London. On May 10, 1886, Secretary Thomas Bayard sent a letter to the British minister in Washington, Sir Lionel Sackville-West, protesting the seizures of both the *David J. Adams* and the *Joseph Story*, in which Bayard stated:

> The seizure of the vessels I have mentioned, and certain published "warnings" purporting to have been issued by the Colonial authorities, would appear to have been made under the supposed delegation of jurisdiction by the Imperial Government of Great Britain, and to be intended to include authority to interpret and enforce the provision of the Treaty of 1818, to which, as I have remarked, the United States and Great Britain are the Contracting Parties, who can alone deal responsibly with questions arising thereunder.

It is clear from these diplomatic exchanges that the United States had no interest in dealing with little "colonial" Canada on this matter. Bayard further warned West that Great Britain and the United States should not let the great international agreement of the Convention of 1818 to "become obscured by partisan advocacy or distorted by the heat of local interests." Bayard made the accusation that Canadian officials had "improperly expanded into an instrument of discord by affecting interests and accomplishing results wholly outside of and contrary to its object and intent," and thus interrupted the legal and happy commercial relationship between American schooners and local fishermen. Bayard concluded his letter with a more forceful warning: "It is obviously essential that the administration of the Laws regulating the Canadian inshore fishing should not be conducted in a punitive and hostile spirit, which can only tend to induce acts of a retaliatory nature."[61]

Edward Phelps further accused the Canadians of an unwarranted and economically biased interpretation of the international agreement by searching for the narrow "technical effect of the words," rather than a "construction most consonant to the dignity, the just interests, and the friendly relations of the sovereign Powers."[62] The American discourse continued along this line in numerous diplomatic communications with London and British representatives in Washington. In these correspondences, Phelps and Bayard continued to accuse "local agents" of aggressive "agitations" or "ill temper" actions against American fishing schooners legally conducting trade in Atlantic Canadian ports.[63]

Canadian officials quickly responded to the American challenge of their jurisdictional authority. They argued that under the provisions of the British North America Act, the Dominion had the right to manage international treaties signed by Great Britain that affected its own territorial possessions. Furthermore, they maintained, the British Crown fully endorsed the Dominion parliament, and they reminded the Americans that the Treaty of 1818 was not a treaty between the United States and the British parliament, but a treaty between the United States and the British monarchy. Under such an interpretation, Canada's Department of Justice concluded that "An Act of that [Canadian] Parliament duly passed, according to constitutional forms, has as much the force of law in Canada, and binds as fully offenders who may come within its jurisdiction, as any Act of the Imperial Parliament."[64] Thus, the British North America Act gave the Canadian parliament, not the British parliament, the right to regulate both the Canadian economy and Canadian resources.[65]

This protection of the Canadian economy did not violate the international agreement of 1818 because, as Canada's Privy Council interpreted in 1886, that treaty intended to create an equitable use of the totality of the North Atlantic fisheries. When the United States closed out Canadian fishermen from the American markets, Canada had the right to restrict the bait trade because the Convention of 1818 was "framed with the object of affording a complete and exclusive definition of the rights and liberties which the fishermen of the United States were thenceforward to enjoy in following their vocation, so far as those rights could be affected by the facilities for access to the shores or waters of the

British provinces, or for intercourse with their people."[66] Allowing Americans to continue to utilize the advantages of Canada's inshore waters via the bait trade, without reciprocity for Canadian fishermen, would create an unfair advantage for Yankee fishermen and thus violate the spirit of the 1818 agreement. The Canadian diplomatic discourse in 1886 thus suggested that a restriction of the bait trade did not violate the 1818 agreement, but quite the contrary, was almost required by the 1818 agreement.

New England fishing interests, particularly those in Gloucester, obviously responded to the Canadian seizures with a great deal of anger. They already had allies in the Republican Party who were eager to strengthen their closed-market policies. The seizures, and particularly some of the more dramatic displays of force by the Canadian Marine Police, tended to turn American public opinion against Canada. The simple matter of purchasing baitfish from fishermen in Nova Scotia suddenly became a serious issue of national pride in the United States. Few people contributed to this transition more than Republican senator William Frye of Maine, who in January 1887 introduced a "Retaliatory Bill" that would authorize the president to seize any Canadian vessel conducting trade in United States ports. In the United States, there was a great debate on the proper response to the Canadian seizures and harassments of American fishing schooners. During the debate, Frye lamented: "Canada is now playing the same role she has twice before; that she plays it for but one purpose, and that to secure negotiations and reciprocal treaties." Frye referred to similar tactics used by British and Canadian marine forces just before the signing of the Reciprocity Treaty of 1854 and the Treaty of Washington of 1871, both of which, according to Frye, created disadvantageous trade patterns for the American domestic economy.[67] Frye called upon the president not to negotiate trade treaties under threat, but instead to close the ports, which would "end all trouble between the United States and Canada; Canada will stop her outrages, and then if negotiations are desirable we can enter into them; if, then, a reciprocal treaty is desired it can be made. It must not be demanded under fire; it will not be granted under the pressure of threats, insults, and outrages."[68] Frye was in the process of turning this little debate over baitfish into a major issue of national honor.

In addition to the seizures, Frye and other New England politicians questioned the more general harassment of American fishermen in Canadian waters. Frye attacked Canada for closing the Strait of Canso between Nova Scotia and Cape Breton, which, Frye argued, was "a public thoroughfare, a highway for all vessels of all nations of the earth." Frye continued his speech by giving a list of other "harassments," which included the boarding and confining of American vessels for landing crew members, hiring pilots, entering the Bay of Chaleurs, and attempting to enter harbors for safety from weather or to purchase food, wood, or water. One of the more often cited examples of these harassments was the story of the Gloucester schooner *Molly Adams*. While off the shore of Prince Edward Island, the *Molly Adams* picked up seventeen crew members and some of the cargo of the Canadian fishing schooner *Neskita*, which had foundered in a storm. Upon entering the bay, the *Molly Adams* came across the Canadian Marine Police schooner *Critic*, which refused the *Molly Adams* the right to land the cargo and initially provided no assistance to the distressed crew of the *Neskita*. According to Frye, the rest of the fishing voyage of the *Molly Adams* was a total loss on account of searching for passage home for the stranded crew and loss of provisions from feeding an extra seventeen men. Frye concluded the story: "Now, I say, Mr. President, that you may search the annals of all history, you may even go to that of the Fiji Islands and you can not find a case which for brutality and inhumanity exceeds that. . . . This took place in the nineteenth century in a civilized and Christian country! The Canadian people are not responsible for it, but the government officials are."[69]

In New England, and throughout the country, the issue became sharply divided down partisan lines. In March 1887, Maine Democrat and later commissioner for the 1888 fisheries treaty William L. Putnam presented a paper to the Portland Fraternity Club outlining the history of the disputed fishing territory in the North Atlantic border seas. Putnam observed the underlying truth of the whole affair when he stated, "As these waters become year in and year out more crowded, with consequent increasing necessity of more and more strict and accurate determination of co-terminus rights, the difficulties coming from such expressions will multiply."[70] According to Putnam, the 1818 treaty worked well enough in 1818, but with more fishermen, who used more advanced gear, operating in the contested waters, a new treaty was essential. Putnam sympathized with the American fishermen, who often found themselves chasing fish from outside the limits to inside the limits, yet "it is not in the nature of the hound to abandon its chase of the fox in order to spare a farmer's cornfield, nor in the nature of the Yankee fishermen to refuse to pursue on the high seas a school of mackerel over an invisible line."[71] According to Putnam, the realities of the North Atlantic fisheries' environment required more open access and regulation of those fisheries by international law for the common good of all nations, and likewise the opening of the fish marketplace to international free trade—something Maine Republican Senator Frye would never agree to.

Putnam sympathized with Nova Scotia fishermen and saw both Canadian and American fishermen as sacrificial pawns in world diplomacy. While the fishery business was but one part of the massive American economy, "to the people of Nova Scotia, living on the southern rock-bound shores of the peninsula, gathering the most meagre livelihood from the scanty soil of their coasts, eked out by what the sea affords, the fisheries are a matter of life." Thus, Putnam suggested that Frye and his fellow Republicans might temper their anger, because "we *might* be generous, we *should* at least be conciliatory and temper 'justice.'"[72] Putnam hoped to remind his audience of "the declaration of many a skipper, that in port after port he had received the kindliest treatment from the inhabitants of Nova Scotia, old-time acquaintances."[73] Putnam was often more hostile to American manufacturing interests, as well as their Republican allies in Washington, who sought the protection of their domestic market at the expense of the livelihood of North Atlantic fishermen, both Americans and Canadians, who relied on larger global markets to stabilize the price of their caught commodities.

Other Democrats, like Frank Lawler of Illinois, questioned the readiness of the United States to face the possibility of war with Canada and Great Britain. On January 31, Lawler called upon the president to report on the conditions of coastal defense in preparation for the 1887 fishing season.[74] By the winter of 1887, many papers throughout the United States began to question the necessity of Frye's proposed retaliation bill. In March 1887, Senate Democrats sought to lessen the bellicose nature of Frye's retaliation bill by allowing the president only to exclude Canadian fish products and not to close United States ports to all Canadian vessels.[75]

By the spring of 1887, rumors circulated that Canada and Britain might relax the enforcement of their strict interpretation of the 1818 agreement. On March 24, Lord Salisbury sent word to Washington that Britain and Canada were prepared to return to the rules and regulations as set by the Treaty of Washington for the upcoming season, and if necessary beyond that, "without any suggestion of pecuniary indemnity."[76] Opposition papers in Nova Scotia were quick to attack those who were "ready to give away our fisheries, for any slight concessions that would be of benefit to them." The *Novascotian* warned that they were "not prepared to have our shore fisheries lightly considered, or our inalienable rights in them cheaply bartered away by the imperial authorities."[77] Although it soon became apparent that it was the imperial authorities in London who ordered the relaxation of enforcement in the hope of negotiating a settlement with Washington, the Liberal press in Nova Scotia continued to attack the Conservative government in Ottawa, who by creating the tariffs under the 1879 National Policy, which affected British exports to Canada as much as it did American exports, so alienated London from the concerns of Canada that it should be no surprise that London now abandoned the Canadian fishermen.[78] Liberal papers and politicians thus accused the Conservative government of quietly giving instructions to the officers of the Canadian Marine Police "not to be so vigorous in the prosecution of their duty as they seem to have been last year."[79]

Although there were occasional disputes concerning the rights of American vessels to use Canadian ports during the fishing season of 1887, the spectacular conflicts of 1886 did not repeat themselves.[80] On May 4, 1887, the Grand Manan schooner *Josie L. Day* was seized for smuggling goods into St. John from Eastport, Maine, but was later released with no charges or fines applied.[81] In June 1887, the *Novascotian* reported American fishing schooners casting seines within one mile of the coast of Cape Breton and hiding their identity by covering their sterns with canvas.[82] In July, the paper again reported the illegal activities of American fishing schooners, this time poaching off Summerside, Prince Edward Island.[83] One Gloucester schooner, the *Colonel J. H. French*, ran into trouble with the Canadian cruiser the *Critic*. On July 25, the *Critic* took two seine boats of the *French* found illegally fishing for mackerel within the limits near East Point, Prince Edward Island.[84] Another Gloucester schooner, the *Annie W. Hodgson*, was fined $400 for not reporting to the customhouse after anchoring near Shelburne, Nova Scotia. Master Morrison of the *Hodgson* claimed he was only in Shelburne in search of a missing dory and its two-man crew.[85] New England papers responded to these affronts with a brief fury, but the situation subsided when the vessel was released upon payment of the fine.[86]

Much of the political discourse in both the United States and Canada tended to blame local biases as the root of the problem. Republicans in the United States attacked Dominion officials and criticized the British for allowing Canadian officials to usurp the right to manage and police the international fisheries of the North Atlantic. This interpretation of a national dynamic in Canadian policy was at the heart of the report issued by the U.S. Committee on Foreign Relations in the United States Senate in 1887. It concluded:

> From all this it would seem that it is the deliberate purpose of the British Government to leave it to the individual discretion of each one of the numerous subordinate magistrates, fishing officers, and customs officers of the Dominion of Canada to seize and bring into port any American vessels, whether fishing or other, that he finds within any harbor in Canada or hovering within Canadian waters. The statute does not even except those Canadian waters in which, along a large part of the southern coast and the whole of the western coast of Newfoundland, they are entitled to fish, to say nothing of the vast extent of the continual coast of Canada.[87]

This debate in the United States most often pointed to a diplomatic solution to the problem. This diplomatic solution, it was argued, could be based on rational interpretation of law, rather than the often irrational behavior of fishermen and the politicians who pandered to fishermen. As such, this logic sought to divorce fishery management from fishermen and local agents and further invest it in international agreement and management.

As secretary of state, Thomas Bayard communicated to Great Britain the popular opinion of the American people and the U.S. government that the British and Canadians violated legal codes of conduct and international treaty stipulations. Yet, as a Democrat and an appointee of President Grover Cleveland, Bayard was certainly not an Anglophobe and often suggested the idea of exchanging free trade for Canadian fish merchants in the United States for free American access to fish resources in Canada. The election of Cleveland in 1884 and the Democratic majority in the House of Representatives in both the 49th and 50th Congresses gave Democrats and their free-trade allies a window of opportunity to restructure the tariff laws and reduce the federal budget surplus by eliminating certain duties. This, the Democrats believed, would encourage the growth of small businesses in an otherwise monopoly-dominated American industrial environment.

While New England, and especially its fishing interests, remained solidly Republican and opposed to free trade, certain groups within the industry, primarily fish buyers and sellers, actively campaigned for free trade and open ports. After Cleveland's inauguration in 1885, the Committee of the Boston Fish Bureau began an active campaign to lift the trade barriers on Canadian fish products. In a circular of September 1885, the committee argued that the tariff drove up the prices of fish products. This greatly benefited large fishing firms and disadvantaged smaller operations, as well as consumers of the products. The circular attacked the larger fishing operations and argued: "The people who will gain anything by the exaction of duties are a few hundred vessel-owners in New England. . . . Is it fair that we should be taxed for their support, or that a few owners of fishing-vessels should reap an advantage obtained at the expense of the great body of consumers of fish in all parts of the country?"[88] This argument mirrored the general historical context of the Democratic platform during the late nineteenth century.

This view, however, was held by the minority throughout New England, a region that continued to oppose the free trade of fish products with Canada. In addition, in both the 49th and 50th Congresses, the Republicans remained in control of the Senate, and therefore they shaped any international agreement related to free trade. In the heated atmosphere of 1886, it would have been impossible for the Democratic majority in the House to achieve free trade with Canada. Public opinion described the aggressive exclusion of American fishing vessels from inshore waters by the Canadian Marine Police as an affront to national pride. This elevated sense of patriotism led to many calls for an equally aggressive campaign against Canadian interests in the American market.[89]

The investigation launched by the Senate Committee on Foreign Relations in 1887 supported William Frye's earlier suggestion that the United States government should respond with exclusionary legislation that would prohibit the free importation of Canadian products. Many in the American government argued that the efforts of the Dominion to exclude the Americans from their fisheries had very little to do with the fisheries themselves, but were designed to force the United States to adopt more liberal trade policies. Republican senator George F. Edmonds of Vermont and Democratic representative Perry Belmont of New York authored the necessary retaliatory legislation. In the meantime, the Committee on Foreign Relations authorized the president to implement at his discretion what became known as the "Retaliation Bill." National pride shaped the rather quick and one-sided debate. Senator William Frye was particularly outspoken in his defense of American pride in opposition to the upstart Canadians:

> Mr. President, the American people know that they are sixty million; they know they are the richest nation on earth; they know their power. They will not humiliate themselves and outrage their conscious pride. . . . [T]he interest and the dignity of this Republic requires [President Cleveland] by

> public proclamation to close our ports to all Canadian fishermen, against any merchant vessel or all merchant vessels, against any of the products of Canada, he shall have the power to do it.[90]

Within the context of this enlarged sense of national pride, the bill passed through Congress with overwhelming approval, thereby leaving President Cleveland in a difficult position. He could not enforce the Retaliation Bill without losing his free-trade allies, yet he certainly could not ignore it after it received such broad support both in Congress and in public.[91]

Through the fall of 1887, pressure mounted in both the United States and Canada to find a solution to the fishery problem. The *Novascotian* reported that the commission to solve the problem "ought to be one of the most important events in Canadian history that has occurred in many years." The paper further warned that "it is of the greatest importance to Canada that the privilege of fishing in her waters, buying bait from her people, and other advantages" for the American fleet "should not be underrated and bartered away for some wholly inadequate concession on the part of the United States."[92]

In Washington, President Cleveland decided not to enforce the Retaliation Bill. Instead, he instructed Bayard to invite Canadian and British representatives to discuss the situation. Prime Minister Macdonald sent his long-time ally Sir Charles Tupper of Nova Scotia to Washington to work alongside the British diplomat Joseph Chamberlain, in order to orchestrate what the Canadian Conservative Party believed were treaty negotiations.[93] Although neither Cleveland nor Bayard would have used the words "treaty negotiations" to describe the talks, they quickly developed into essentially that.

The idea that the executive would have launched such negotiations in direct opposition to the clearly stated goals of the Senate would have led to a political disaster for Cleveland. Bayard kept this point in mind as he continued the talks with Tupper and Chamberlain. These talks resulted in the proposed Treaty of Washington of 1888. This treaty outlined what many in the United States believed were more restrictive fishing rights for the Americans than those provided by the Treaty of 1818. The treaty also articulated in specific language the philosophy that Canada had voiced the previous two fishing seasons, one that gave the Dominion the right to restrict American activities along the coasts and in the harbors of the Maritime Provinces. It also authorized American fishing operations to use the Strait of Canso and other bays, which many Americans believed was a right guaranteed by international maritime civility and did not require the consent of Canada. It prohibited many activities in Canadian ports that New England fishermen believed the free port agreements of the 1830s had previously granted, and the treaty language did little to clarify the contested terms of the Treaty of 1818. Fishing interests in the United States believed that the resulting Treaty of 1888 contained far too many dangerous precedents that could later be used by Canada to further restrict American fishing interests.[94]

While the proposed Treaty of 1888 contained numerous technical flaws, the debate over ratification in the Senate ignored most of these and instead took on a decidedly political tone. The Republican-controlled Committee on Foreign Relations immediately attacked President Cleveland and his Democratic allies in Congress for negotiating a treaty without the clear consent of the Senate. The committee argued that Cleveland's action not only violated the terms of the Constitution, but was dangerous to national security. It concluded that "It is not difficult to see that, in evil times, when the President of the United States may be under influence of foreign and adverse interests, such a course of procedure might result in great disaster to the interests and even the safety of our Government and people."[95] The strong language used by the committee in this report suggested that the president actively undermined the security of the United States and violated the principles of a representative government. Considering the great labor unrests of 1886, the financial crises, and the rising critique of political corruption in the United States, it is not surprising that this little fishery treaty became a casualty of larger political issues.

The president's allies in the Senate did their best to defend him and Bayard by emphasizing their right to direct the nation's diplomatic efforts. Senate Democrats portrayed their Republican counterparts as the real usurpers of constitutional powers by arguing that their legislative role did not necessarily supersede the executive's power. In the minority report, the Democrats characterized the Republicans as stubborn, self-righteous men who did not wish to play by civilized rules of diplomacy, but instead sought to bully their way through international politics. Furthermore, they feared that the Republican policy would only lead to violence if not outright war with Great Britain. The Democrats attacked the Republicans in their own report, stating, "They prefer the chances of greater success through legislation that will intimidate the British Government or greatly embarrass British commerce. This seems to indicate that they rely for success more upon British cupidity and the fear that Government has of the consequences of war, than upon its sense of justice, or its good faith in keeping treaty obligations."[96] Fear of violent conflict with Canada or Britain remained a very real concern for some in Washington, as both the United States and Great Britain had sent naval forces to the North Atlantic to oversee the fisheries dispute.

These Democratic senators argued that Bayard and Cleveland had successfully ended the nation's most daunting challenge to international peace. They retold the long history of the dispute by emphasizing the numerous moments in which the nations came to the brink of war. For example, they concluded:

> Under the misunderstandings of the past we have on both sides sent fleets to these waters to protect our fishermen against each other and against the unfriendly conduct of the local governments; fleets to enforce agreements that the governments concerned could not expound by a mutual understanding. . . . If these questions are left open, and commercial war is inaugurated through measures of retaliation, how many ships and guns is it supposed will be needed to keep the peace between our fishermen on the coasts of Labrador and Newfoundland?[97]

In this interpretation, Bayard and Cleveland had solved an international dispute that could have easily led the nation down the path to conflict. The minority report expressed the growing faith in international arbitration and diplomacy to avoid direct conflict between economic or popular impulses.

This philosophy maintained that neither diplomats nor angered national governments produced the danger in the North Atlantic fisheries. The Democrats in the United States Senate argued that the constant contact and regional rivalries among fishermen and fishing businesses created the real agents for the difficulty in the international fisheries. Cleveland and Bayard hoped to avoid this conflict by opening both the markets and the waters to complete use by all respective parties. Only by eliminating regional control could the fisheries be managed in a successful and peaceful manner. The minority report continued in this vein and stated that

> The danger in this direction does not come from the desire of either Government to promote a war, but from their inability to prevent its initiation through the personal hostilities of men associated in the use of common rights and privileges, and stimulated by rivalries which are encouraged by

> laws of retaliation enacted by their Governments. . . . These are some of the dangers against which this treaty wisely makes safe provision.[98]

Policymakers became increasingly fearful of the individual impulses of fishermen upon the fishing grounds of the North Atlantic, and the Democrats vociferously argued that free trade created the context in which peace would prosper. While the context of labor unrest in the late 1880s certainly explains some of these fears, the specific history of the fisheries dispute shows that the fishermen were never the ones calling for aggression of stringent enforcement—that blame clearly falls on radical politicians and their newspapers.

In Canada, Sir Charles Tupper echoed the Democratic fear that the present situation of exclusion and American retaliation could lead to a private war between the fishermen. In his official report to the parliament, Tupper argued that the "fishermen, perhaps, are the most intractable and uncontrollable people in the world, and when a fisherman gets on board his little smack he thinks he is a monarch of all he surveys, and he can go where he pleases, and do what he pleases. The result was that, as before, collisions occurred."[99] Supporters of the treaty in both the United States and Canada suggested that the high level of regulation embedded in the compact was necessary to avoid regional disputes about fishing groups.

The Canadian opposition to the treaty, mainly the Liberal Party and its press, continued its critique of the Conservatives' abandonment of North American free trade in their hope of constructing a national economy revolving around central Canada. The treaty of 1888, the *Novascotian* recognized, created a highly regulated international fishery economy without providing for the Liberals' primary agenda: North American free trade. The paper recognized that the treaty's prohibition on American inshore fishing in Canada caused no significant sacrifice for the New England fishing industry. The treaty did, however, give to the Americans their primary goal of "having Nova Scotia for a base of operation," without an equal exchange of Canadian free access to the American market.[100] The Liberals in Nova Scotia were not opposed to the presence of Americans in their ports and waters; in fact, they encouraged the increased trade that came with the American customers. They only sought equally free access to American markets, which the Treaty of 1888 did not provide.

In the United States, the Senate debate over ratification resulted in numerous public speeches and letters from notable statesmen and lawyers. New York lawyer Charles Davison argued that the treaty language supported Canada's claim to "ownership" of the waters. To accept the treaty would result in the acceptance of the Canadian claim to jurisdictional authority and thus bring an end to the long-held belief in the United States that only the governments of the United States and Great Britain could claim jurisdictional authority. Lawyer Joseph Doran echoed Davison's fears that the fisheries would become private property, stating that "the fundamental principles of property and sovereignty, which are the basis of our fishery rights, seem to have been abandoned at the outset by our own commissioners and our case to have been presented and considered, as Mr. Bayard expressed it in his letter to a gentlemen in Boston, on the basis of the 'Fishery rights of one country in the jurisdiction of another.'"[101] According to these American legal scholars, the Treaty of 1888 would result in the removal of fishery management from international debate and arbitration and put it solely in the hands of Canada and its provincial governments—the very source, they argued, of all the problems.

Cleveland continued to defend the treaty, claiming that it was framed in the "spirit of liberal equity and reciprocal benefits," and that its acceptance by the Senate could be the "only permanent foundation of peace and friendship between [Nation-]States."[102] Yet the president received little support outside of his Democratic allies; few in the fishing industry were satisfied with the negotiations. The secretary of the National Fisheries Association, Luther Maddock, publicly condemned the treaty for extending to American fishermen rights that he argued did not require treaty arrangement but were understood as measures of international civility. For example, Maddock particularly attacked the treaty for explicitly granting to Americans the right to land at ports in case of damage, illness, or poor weather. This right, he maintained, was widely considered universal and "humane privileges, such as all civilized or semi-barbarous nations grant to each other, even when there is no treaty to bind them."[103] Thus, it should not be included in the treaty provisions.

Although the Republican-controlled Senate defeated the proposed Treaty of 1888, one element of the negotiations remained intact. In an attempt to secure American favor during the negotiations, the Canadian government adopted a modus vivendi. This policy allowed American fishing interests free access to Canadian ports with the purchase of a license amounting to $1.50 per ton. This measure, Tupper argued in front of the Canadian parliament, would convince the American public of the necessity of free access to Canadian ports and thereby eventually lead to the acceptance of a permanent exchange of these rights for free Canadian access to American markets.[104] The modus vivendi was a legislative act of the Canadian parliament and thus went into effect even with the American defeat of the Treaty of 1888.[105] Beginning in the summer of 1888, New England schooners had free access to Canadian bait, supplies, and crew with the purchase of a relatively inexpensive license, and the seizures and hostilities of 1886 came to an end.[106] Factions in both countries continued to push for free trade, but the idea of protected domestic markets remained strong in the United States, and free trade would not be a serious topic of conversation until the administrations of William Howard Taft and Wilfrid Laurier and their proposed trade agreement of 1911. That reciprocity treaty also failed, and the United States and Canada would not sign a comprehensive North American free-trade agreement until the late twentieth century.[107] The Canadians essentially won the legal debate following the Vice-Admiralty Court's decision in the *David J. Adams* case. When the court condemned the *Adams*, it essentially agreed with the Canadians' claim of territorial authority over the inshore waters of Canada and any American fishermen seeking business in those waters. Yet after the failure of the 1888 treaty, the Ottawa government politically abandoned that legal victory by extending the modus vivendi of 1887, which opened the waters to American commerce without securing free markets for Canadians in the United States.

Even in the absence of a free-trade agreement, Canada continued the modus vivendi of 1887 and the licensing system that allowed American fishermen into Maritime Canada. This policy leads to some important conclusions concerning the purpose of Canadian fishery management in the late nineteenth century. Although the Dominion put up a strong front and aggressively pursued a policy of exclusion in the late 1880s, the language of their diplomatic correspondences and their adoption of the modus vivendi of 1888 clearly illustrate that their real agenda was not to protect the Canadian fishermen or Canadian fish for exclusive local use. Instead, the government wanted to secure better trade terms with the United States, principally the free entrance of Canadian-caught fish into the American market. When it became clear that these terms would not be obtained, regardless of Canada's aggressive exclusionary measures, the Dominion gave up its attempt to exclude American fishermen from their bait supply by continuing the modus vivendi of 1888 for several decades. It is clear that the Canadian government held no real interest in preserving local control of a local resource. In fact, it used that local resource to secure national goals, which was the entire purpose of

Macdonald's National Policy. Thus, the Canadian government used the Maritime bait supply as a bargaining device in a national strategy for free trade.

This larger diplomatic and political history is well documented. On both sides of the border, liberals and conservatives seemed to place the blame for the growing hostilities upon the fishermen who, according to the diplomats, politicians, and newspapers, fought endlessly for various unexplainable reasons. Within this well-documented context, it is difficult to understand the realities of those local fishermen who, at times, eagerly sought out commercial relations with American schooners illegally plying the domestic waters of Atlantic Canada. We may often forget that the *David J. Adams* was in Annapolis Basin because the captain knew he could easily obtain fresh baitfish from the local fishermen despite international and national laws that prohibited this trade. The trial records of the *David J. Adams* provide historians with a rare opportunity to read the thoughts of those local fishermen, to learn why they sought out illegal trade relations with the Americans, and to understand how this exchange occurred. When compared with the political documentation, we learn that local Canadian fishermen and migratory American schooners created a complex set of informal codes of conduct to regulate their relationship, which were outside the bounds of the formal legal and diplomatic relations.

Smuggling fish in and out of the British possession in the Atlantic was nothing new in 1886. In fact, for most of the region's history there was not even any smuggling. The law regulating commerce in the North Atlantic fisheries changed several times throughout the nineteenth century. The Convention of 1818, the Reciprocity Treaty of 1854 lasting until 1866, and the Treaty of Washington of 1871 lasting until 1883 all stipulated different terms for fishing and fishing commerce. At times, selling bait was a legal act of trade; at others, it was smuggling. Yet throughout, local fishermen kept doing what they had always done: they traded fish with the Americans. Although the nation-states changed the language of this commercial activity from "trading" to "smuggling," the fishermen changed few of their habits. When we look at it from this perspective, it really does not seem all that shocking that a few Digby fishermen sold baitfish to the *David J. Adams.* What were they really doing but continuing a practice that had existed since the first day that a New England schooner arrived off the coast of an Atlantic Canadian town? They traded fish.[108]

George Vroom, a fisherman and farmer from Clements, a town in the Annapolis Basin just outside of Digby, testified that he had sold four-and-a-half barrels of herring and two tons of ice to the *David J. Adams*. Vroom stated that the captain of the American schooner wanted twenty barrels of herring and would pay as high as a dollar per barrel. Vroom alleged that the whole deal was made while the captain was ashore, during which time, Vroom stated, "He talked about the treaty, and I said I could see where his vessel came from; he replied yes, she was an American vessel; that he could get his supplies from New Brunswick, have his crew and carry his fish to Gloucester, pay the duty, and make more money than he could by fitting out in Gloucester, because he could hire his men cheaper and buy his supplies cheaper."[109]

Robert Spurr, another weir fisherman from Clements, stated that he was unable to see the name or port of registry of the *David J. Adams*, and further testified that the captain claimed it was the *Adam Craft* from North Shore.[110] Samuel D. Ellis, a Victoria fisherman for fifty years, also claimed that Master Alden Kinney first attempted to deceive the local fishermen concerning his nationality. Ellis stated that when Kinney came ashore looking for herring, "I told him no, because it was against the law, and that I could not sell to Americans. He replied that the schooner had been American, but the English had bought her." Ellis did not believe Kinney's ploy, and both knew full well where the *David J. Adams* was really from. Ellis, however, continued the bargaining, later stating, "He asked me what the price of the bait was, and I told him it was one dollar a barrel. He said he would give me one dollar and twenty-five cents a barrel for it if I would let him have it. He offered me one dollar and twenty-five cents, and I took it." Ellis's respect for the law, and possibly even his nationalism, waned in the face of the high value offered by Kinney and perhaps other American fishing masters in Nova Scotia. It appears that capitalism trumped nationalism in this incident.[111]

But one question still persists: why was the *David J. Adams* in Digby, Nova Scotia—a port on the Bay of Fundy—looking for baitfish for a voyage to the Grand Banks? Why would Master Alden Kinney go so far as Digby when baitfish could have been just as easily obtained on the southern coast of Nova Scotia, more on a direct line to the Grand Banks? Perhaps the answer has to do with Kinney himself. Alden Kinney, the master of the American fishing schooner, was originally from New Brunswick. He had most likely stopped by his old home on his way to the fishing grounds, as visiting relatives was a common practice among American fishermen.[112] If so, then Digby, Nova Scotia, was just a short distance away, and perhaps Kinney already had contacts in Digby from his days in New

Brunswick. Kinney may also have been on the north side of Nova Scotia looking for crew. The *Digby Weekly Courier* reported that six of the eighteen fishermen aboard the *David J. Adams* were Canadians, while the *Morning Herald* of Halifax provided a full crew list, which identified four men from St. George, New Brunswick; one man from Shelburne, Nova Scotia; and one man from Pubnico, Nova Scotia—while the rest of the crew were from Maine, Sweden, Germany, or unknown. Not a single crew member was from Gloucester, Massachusetts.[113] This points to a key element of the American-Canadian bait trade: it was conducted primarily by Canadian fishermen working aboard American schooners as both fishermen and agents who could find supplies among their own neighbors—a pattern well entrenched in the North Atlantic fishing industry by the end of the nineteenth century. On June 25, 1886, the *Digby Weekly Courier* summarized the importance of local fishermen conducting this trade: "No doubt, the fact that the greater part of their fleet is manned by men from our own shores, has been the reason why they have been able to obtain it [bait] at all, and now that the coast is so well guarded, by our cruisers, this last resource is likely to fail them."[114]

At the end of the fishing season, Captain Peter A. Scott called the American fishermen "without exception the most infernal thieves I ever came across," and said that they "have resorted to every little trick and dodge conceivable" to steal Canadian property. Yet, in his pre–fishing season interview with the press of Nova Scotia, Scott responded to a question about Canadian crew members of American fishing schooners, saying, "Yes a good many; plenty of Nova Scotians and men from Cape Breton, some of them 'whitewashed' by a brief residence on the other side."[115] There is abundant evidence from both the Canadian and American press that many of those Yankee "infernal thieves" were in fact Canadians.

On April 2, 1886, the *Digby Weekly Courier* noted that the American schooner then in Pubnico was mastered by a man from Argyle, Nova Scotia, and that he and his crew, who were also primarily Canadian, regretted the absence of a treaty that had previously made their job much easier. The *Courier* further implied that it was not the fishermen who manned the New England fleet that opposed a free trade for free fishing treaty, but it was the "New England owners and outfitters who are really opposed to the treaty."[116] In the United States press, the commercial newspaper *Broadstreet* noted that "a large majority of the men on American vessels are British subjects, so that the interests of very few persons except the vessel-owners can be served by protective duties."[117] This coverage suggests, contrary to the stated opinions of men like Charles Tupper and Thomas Bayard, that it was doubtful that the actual laborers of New England's fleet generated the difficulties of 1886, but instead many of these fishermen retained close connections with their Canadian counterparts and with the numerous inshore fishermen who provided them with bait.

When the Dominion government sought to prohibit American fishing vessels from entering Canadian ports to purchase this supply of bait, it likewise restricted the ability of American schooners to hire local crew. On March 30, a representative of the fishing firm Hodgkins & Sons of Ellsworth, Maine, was in Cape Sable making arrangements "to have the men taken across to Maine in Nova Scotia vessels" so as to work around the new Dominion restrictions. The firm chartered a schooner specifically for that task and had already hired forty men from Cape Sable alone.[118] Many other firms followed the same practice. Fishing schooners from Boothbay Harbor arrived in Pubnico in late March 1886 and hired ninety men from "different points in Yarmouth and Shelburne counties," who were then transported aboard a steamer to Maine at an added cost of $500 to $700.[119] The area around Yarmouth and Pubnico was the primary geography for this shipment of crews. The *Halifax Chronicle* reported that "A number of fishermen had, as usual, shipped from here [Pubnico] to go in Gloucester vessels, they to call for them on their way to the Banks."[120] The *Novascotian* likewise reported that the Canadian steamer *Lansdowne* warned off several Yankee schooners, "with Pubnico captains," from West Pubnico. The schooners were there to "ship their crews. . . . Hundreds of men have been shipped every year by Yankee schooner between Yarmouth and Cape Sable."[121] In their own general opposition to the 1886 Canadian policy of restricting American access to Canadian ports, the *New York Herald* noted that such a policy seriously hurt the local economy of Nova Scotia, not only because of the loss of the valuable bait trade, but also because most of the American fishing fleet "engaged the services of Canadian fishermen to serve on their vessels," who now found it more difficult to find suitable employment.[122]

Canadians were eager to work for the American fleet not only because of the higher wages, but especially because of the free access to the American market as both sellers of fish and buyers of commodities. In June, at the end of the fishing season, the Gloucester fishing schooner *Plymouth Rock*, mastered and manned by men from Pubnico, Nova Scotia, noted their dissatisfaction with the present situation mainly because they were unable to purchase American goods while in Boston, for fear of having those goods seized upon return to their homes in Pubnico. The master noted that he and his crew had purchased $1,500 worth of supplies the previous year, but now had to turn to the more expensive market in Halifax for their home needs.[123] Following the troubled years of 1885–1887 and with the 1887 modus vivendi in force, American schooners could return to their policy of hiring crew members in Canada. In March 1889, the *Digby Weekly Courier* reported that fishermen in Shelburne, Nova Scotia, the traditional port for labor recruitment by American fishing vessels, received letters from American captains stating that they "need not proceed to the United States this spring to obtain employment in the Yankee fishing fleet, as American vessels will call at Nova Scotia ports and ship men from their own homes."[124] These Canadian fishermen could, once again, work aboard Yankee schooners for American cash, purchase American goods in American markets with the American cash, and have it all delivered to their homes back in Nova Scotia—that is, if they did not wish to remain in the United States following their employment. Despite continued efforts in both Canada and the United States to separate the fish and the fishermen in those nations, both continued to migrate relatively freely.

The year 1886 offers historians a rare opportunity to examine the American-Canadian bait trade that, for the most part, remained beneath the surface of the extensive historical documentation of the profitable and controversial North Atlantic fisheries. Only when that bait trade emerged as an important political and diplomatic topic in Canada, the United States, and Great Britain did it appear in the documentation. Yet, this evidence should not suggest that the trade emerged as a controversy among fishermen in 1886. In fact, fishermen in the North Atlantic constructed a trade system that changed very little throughout the nineteenth century. American fishermen were content with abandoning the inshore waters of Canada for the direct purpose of fishing so long as they retained the right to purchase supplies, hire crews, and transship their cargoes to their home markets. Local Canadian fishermen were often happy to see their American neighbors, for they constituted their principal customers—so long as those neighbors did not attempt to catch their own baitfish in inshore waters. With this informal code of conduct well established by the middle decades of the nineteenth century, cooperation and compromise was the general feeling between American and Canadian fishermen in Nova Scotia, New Brunswick, and Prince Edward Island. Despite what Charles Tupper and Thomas Bayard suggested during those heated debates of 1886–1888, fishermen in the North

Atlantic were not the source of international agitation. Quite the contrary: they had previously established a general system of order resting on an international commodity trade and cross-border employment. In fact, it was the partisan press in the United States and Canada, as well as the elevated sense of national pride rampant in both countries' diplomatic core, that generated ill feelings and an abundant amount of sword-rattling.[125] Thankfully, war was avoided, and the fisheries once again became peaceful diplomatic ground. Yet, throughout the crisis, the fishermen changed few of their habits and retained their locally constructed codes of conduct, which often did the real work in fisheries management throughout the nineteenth century.

"Peaceable Settlement"

THE PROTRACTED DEBATES surrounding the repeal of the 1871 Treaty of Washington and the failed 1888 treaty clearly demonstrated to many that baitfish had emerged as a complex diplomatic issue in Anglo-American relations. The United States Navy, the Royal Navy, and the Canadian Marine Police all sent armed vessels to enforce their own interpretations concerning the rights of American fishermen to purchase bait and supplies, hire crew, and transship cargo in Canadian and British ports in the northwest Atlantic. Politicians and diplomats attempted to solve the issue through legislation, treaties, modi vivendi, and licenses, to little avail. These methods failed because local fishermen sought to construct and enforce their own understanding of resource regulations. Contrary to many national laws and international agreements, these informal codes of conduct stipulated that only local fishermen could extract baitfish from local waters, yet all outsiders, regardless of nationality, could purchase their bait needs from local suppliers. The bait fisheries operated according to these informal codes well enough so long as formal law and lawmakers did not intervene. By the twentieth century, however, a new force in international law crept into the debate and sought to impose the order of legal phraseology on the seemingly chaotic world of informality in the bait business. Unlike previous attempts to control the trade via national politics, this new emphasis on international law proved more successful.

Following the failure of the Treaty of 1888, the Canadian government opened the North Atlantic inshore fisheries to American fishermen through a modus vivendi that lasted for over twenty years. The Canadian government again extended the measure following the failed reciprocity treaty of 1911. The Dominion government, for the most part, sacrificed local rights to the fisheries to appease international interests and to orchestrate the national goal of increased commercialization and industrialization of the Canadian economy.[1] In Newfoundland, however, the colonial government continued to defend its territorial claims and its rights to legislate the local bait fishery. Newfoundland fishermen made good profits by selling baitfish to Gloucester schooners that utilized the deep-sea bank fishery. The relationship between the two groups of fishermen worked out well enough, as long as the Americans continued to buy their bait from local inshore fishermen instead of trying to catch their own. With only a few exceptions, the American fishing industry accepted the terms of this informal code of conduct.[2]

In an attempt to control this trafficking and to increase colonial revenue, Sir Robert Thorburn's Reform Party passed legislation in 1886 that would prevent the sale of bait without a license.[3] This policy intended to direct the bait sale, which operated informally between American schooners and individual Newfoundland fishermen, into the hands of the fish merchants in St. John's. The initiative received sharp criticism from William Whiteway's conservatives, who represented the interests of manufacturers and industrialists and campaigned for the diversification of Newfoundland's economy, including the completion of a railroad that would span the island, which began construction in 1881.[4] Merchant and manufacturing interests thus divided Newfoundland's politics, which added to the heated and sometimes violent debate between the island's Catholic and Protestant population. Within this context of the debate the opposition party, led by William Whiteway, convinced London that Thorburn and his fishing-license program did not represent the majority interests in Newfoundland. As a result, London refused to ratify the legislation.[5]

In 1889 Whiteway's new Liberal Party won office. Under the direction of Robert Bond, the colonial secretary of Newfoundland, the colony began a series of talks with the United States government in an attempt to gain access to the American market duty-free, a goal that had shaped both Newfoundland and Canadian fishery policies since 1854. In 1890 Bond negotiated a treaty with United

States secretary of state James Blaine that exchanged fishing and trading rights in Newfoundland for an open market in the United States. Canada insisted that any talks on the baitfish trade in North America must include the interests of the Maritime Provinces. As a longtime opponent of Newfoundland confederation with Canada, Bond saw no need to inform any officials in Ottawa or any other Canadian governing body of his negotiations with Blaine in Washington. Having thus been excluded from these important trade talks, Ottawa pressured London officials to reject the proposed Blaine-Bond Treaty of 1890. The London government refused to approve Bond's treaty until he gained the support of Canada. In response to Canada's sabotage of his treaty with the United States, Bond imposed restrictions on the Canadian baitfish trade in Newfoundland. This did not endear him to the fishermen of his own colony, and it also further antagonized the Newfoundland-Canadian relationship, which ultimately led to the failure of his 1890 treaty with Blaine.[6]

Ten years later, Sir Robert Bond, now the premier of Newfoundland, negotiated similar terms with U.S. secretary of state John Hay. Bond had learned from his 1890 mistake, and this time he secured supporters in London before going to the Americans. As a result, the Canadian argument for dependent trade agreements between Canada, Newfoundland, and the United States did not attract much attention or respect in London.[7] Therefore, London readily agreed with Bond's proposal and granted Newfoundland the right to sign the treaty with the United States. The 1902 Hay-Bond Treaty, however, failed ratification in the United States Senate after Massachusetts senator Henry Cabot Lodge launched an aggressive campaign to scuttle the agreement, first through delay and then by tacking on so many amendments that Newfoundland could not accept the revised version. Lodge's opposition to the Hay-Bond Treaty reflected the more general Republican opposition to tariff reductions or free trade rather than any specific issue related to the fisheries.[8]

After repeated failures to come to some official agreement with the United States government regarding the baitfish trade in Newfoundland, Premier Bond revised the old Foreign Fishing Vessel Act in 1905. The new measures prohibited foreign fishermen from hiring Newfoundlanders as crewmen, or purchasing bait within Newfoundland's three-mile jurisdiction. Although French and Canadian fishermen still plagued Newfoundland policymakers, Bond clearly stated that this proposal was specifically directed against American interests: "This Bill is framed specifically to prevent the American fishermen from coming into the bays, harbours, and creeks, of the coast of Newfoundland for the purpose of obtaining herring, caplin, and squid for fishing purposes."[9] Bond continued:

> This communication is important evidence as to the value of the position we occupy as mistress of the northern seas so far as the fisheries are concerned. Herein was evidence that it is within the power of the Legislature of this Colony to make or mar our competitors to the North Atlantic fisheries. Here was evidence that by refusing or restricting the necessary bait supply, we can bring our foreign competitors to realize their dependency upon us. One of the objects of this legislation is to bring the fishing interests of Gloucester and New England to a realization of their dependence upon the bait supplies of this Colony. No measure could have been devised having more clearly for its object the conserving, safeguarding, and protecting of the interests of those concerned in the fisheries of this Colony.[10]

Clearly, Bond set out to disrupt American fishing efforts in the North Atlantic regardless of any international agreement. Although Bond intended to force the Americans into negotiations, he effectively destroyed the informal bait trade that existed between Newfoundland's fishermen and American schooners. As a result, he lost the support of the fishing population on the west coast of the island, where this trade dominated the local economy.

American secretary of state Elihu Root communicated to British officials that this colonial legislation represented an objective of the Newfoundland government, based on a local economic bias that prevented the fair and free access to the resources of the North Atlantic, which should otherwise be considered a great international common property. Root argued that only through international management and guidance from Great Britain and the United States could the North Atlantic fisheries remain a free and common resource as intended by the Convention of 1818. In a letter to Sir H. M. Durand, the British ambassador in Washington, he stated:

> I am confident that we can reach a clear understanding regarding those rights and the essential conditions of their exercise, and that a statement of this understanding to the Newfoundland Government, for the guidance of its officials on the one hand, and to our American fishermen for their guidance on the other, will prevent causeless injury and possible disturbance,

> such as have been cause for regret in the past history of the north-eastern fisheries.[11]

Officials in both the United States and Great Britain sought to reconcile their historical disputes in North America by removing control from local governing bodies.

London restructured their imperial interests in North America in order to focus on more pressing global issues, often with the result of appeasing the Americans. The Franco-Russian entente in 1893 forced Britain to reevaluate its naval strategy and contributed to its decision to accept American interests in the construction and control of a canal in Central America, which essentially meant British naval inferiority in the Western Hemisphere. The pressing Alaskan boundary dispute was consigned to arbitration, and the 1903 agreement did little to uphold Canadian interests in the Pacific Northwest. In April 1908, Elihu Root and James Bryce ended the Bering Sea pelagic-seal controversy, and two years later, in 1910, the Passamaquoddy Bay border dispute came to an end when Bryce signed an agreement with President Howard Taft's secretary of state, Philander C. Knox. In many respects, the British eagerly removed themselves as a premier power in North America. This process led one predominant Canadian historian to claim, "The 'slate-cleaning' program of 1906 was an Anglo-American project initiated by Grey and Root and advanced by Bryce; in general, Canada was simply a willing concurrent in most of the treaties that carried it out."[12]

In this diplomatic context, officials in London therefore refused to give consent to Bond's restrictive fishing legislation for fear of upsetting their desired balance with the United States. Instead London, eager to appease American interests, forced the Newfoundland government to accept a modus vivendi that allowed American fishermen to continue their trading relations with local fishermen. At first Bond refused to accept London's modus vivendi. Yet he received little support from the local fishermen, who remained eager to work with the American fishermen, provided that they continue to follow the unwritten local custom of not fishing for their own bait but relying on local fishermen to catch it for them. In a telegram on September 23, 1907, Lord Elgin gave, in no uncertain terms, instructions to Premier Bond that he must accept London's decision. Elgin observed that "His Majesty's Government have received with great regret the refusal of your Minister to cooperate in carrying out the *modus vivendi*, which leaves His Majesty's Government no alternative but definitely to instruct you to publish the Order in Council. This step should, therefore, be taken at once."[13] With that instruction, London officials virtually prohibited the Newfoundland government from any independent action towards the United States, even though only three days later, the British Crown granted Newfoundland independent Dominion status.[14]

The debate concerning the rights of Americans to enter territorial waters in order to conduct the baitfish trade continued well after the 1907 modus vivendi. This agreement only prevented local police forces from seizing American schooners while the diplomats continued to talk. It became clear, however, that the differences in opinion were so extensive that diplomatic communications had little hope of solving a debate that had now lasted for nearly a century. As a result, United States secretary of state Elihu Root and British secretary of state for foreign affairs Sir Edward Grey agreed on January 27, 1909, to send the dispute to the International Court of Arbitration at The Hague. The legal debate at The Hague would focus on the rights of equal use of a common property: the vast sea fisheries of the North Atlantic that had been a geography of international exploitation since their discovery by European powers. Managing the North Atlantic fishery resource would take on a new meaning. Political debates within Canada and the United States became of secondary importance to the legal debate. This debate, orchestrated by eminent lawyers such as Elihu Root, became structured strictly in terms of legal rights of access to common property and the precedent of legal phraseology.

On June 1, 1910, British, Canadian, Newfoundland, and American legal representatives met at the International Court of Arbitration at The Hague for the North Atlantic Coast Fisheries Arbitration Tribunal in an attempt to settle this century-long dispute. In his closing argument for the American case, Elihu Root identified the underlying problem in solving this issue. He stated metaphorically:

> Words are like those insects that make their color from their surrounding. Half of the misunderstandings in this world come from the fact that the words that are spoken or written are conditioned in the mind that gives them forth by one set of thoughts and ideas, and they are conditioned in the mind of the hearer or reader by another set of thoughts and ideas, and even the simplest forms of expression are frequently quite open to mistake, unless the hearer or reader can get some idea of what were the conditions in the brain from which the words come.[15]

Unlike fishermen who handled fish, these international lawyers focused on language. For them, the language of treaties, correspondences, agreements, and modi vivendi guided the way to a stable understanding of the accepted rights and regulations that would manage the fisheries for all those competing factions that exploited this common property. These lawyers tried to use words to create order out of the chaos of competing fishing groups and nationalism, out of the uncontrollable nature of the fisheries, and out of the vastness of the open ocean.

The idea of using arbitration to decide important international legal disputes became what might be considered a legal fad during the late nineteenth and early twentieth centuries. The International Court of Arbitration at The Hague was established by a series of conventions throughout the period between 1899 and 1908; it was widely applauded as a vehicle by which all future international troubles could be solved through peaceful debate. Elihu Root best summed up the ideals of this court in his address at the laying of the cornerstone of the Pan American building on May 11, 1908: "There are no international controversies so serious that they cannot be settled peaceably if both parties really desire peaceable settlement, while there are few causes of dispute so trifling that they cannot be made the occasion of war if either party really desires war. The matters in dispute between nations are nothing; the spirit which deals with them is everything."[16] Today, many might consider something like North Atlantic baitfish as "trifling"; Root knew that it was much more "serious" than that.

Thus the fishery dispute shifted to The Hague, a forum where legal discourse constructed an acceptable decision regarding the managerial rights in the international commons of the North Atlantic. In the end, both the British and the Americans accepted the decision of the court, and many considered the decision one of the greatest victories in the pursuit of international peace. In his opening remarks, the presiding president of the court, Austria's Dr. H. Lammasch, stated that "Every sentence rendered by this Court ought to be by virtue of its impartiality and equity a new marble pillar to sustain the ideal palace of Justice and Peace."[17] In December 1910, President William Howard Taft praised this fisheries tribunal, along with the Geneva settlement of the *Alabama* controversy and the Paris settlement of the Alaskan pelagic-sealing debate as "three great substantial steps towards permanent peace, three facts accomplished that have done more for the cause of peace than anything else in history."[18] Few legal scholars of the day doubted that the logic of law could not solve the great irrational disputes of man caused by local bias and ignorance.

Article 1 of the Convention of 1818, the very same article that had caused political controversy, diplomatic confusion, and fishermen violence ever since its adoption, became the primary focus of the debate. In brief, Article 1 stated that the inhabitants of the United States should have forever, "in common with British subjects . . . the liberty to take fish of all kinds" along the assigned treaty coast, and the right to enter bays and harbors for the express purpose of obtaining water, wood, shelter, or repairs. The key terms in this legal debate at The Hague were "liberty" and "in common with." This was the language that shaped the legal debate concerning the international management of the North Atlantic fisheries.

The first question, considered the most important in the arbitration, specifically addressed the right of British or colonial governments to write legislation regarding the management of the bait fisheries within their three-mile territorial jurisdiction, even if such limitations affected American treaty rights. Or, as the Americans asked, would such legislation, due to its effect on American fishermen, require the consent of the government of the United States? In other words, was the North Atlantic fishery an international common property that would be subject only to international management, or a regional common property subject to local legislation? For the lawyers present at The Hague, the linguistic argument concerning the meaning of the terms "liberty" and "in common with" would yield answers to this century-long debate.

In regard to question 1, the tribunal decided that the waters within the three-mile territorial limits of the British, Canadian, and colonial governments were subject to the legislative control of those governments. American fishermen who claimed rights within those waters must obey such legislation as deemed necessary by the British, Canadian, or colonial governments. However, the tribunal specifically stated that such legislation should and must be "necessary" and "reasonable" for the protection and preservation of the North Atlantic fisheries. Moreover, it could not directly interfere with American treaty rights or otherwise protect exclusive fishing rights for British subjects at the expense of American citizens. Such legislation would violate the "in common with" clause of Article 1 of the Convention of 1818. Furthermore, the tribunal granted to the United States the right to appeal to a joint commission, also established by the court, any legislation passed by British, Canadian, or colonial authorities that the United States believed unfairly favored British subjects over American citizens, or might not be necessary or reasonable for the protection and preservation of the fisheries.[19]

The American government used this appeal process to effectively eliminate the Newfoundland Foreign Fishing Vessel Act. In some sense, the court determined that the North Atlantic fishery was in fact an international common property; it was subject to international management. Although local legislation could control the basic extractive operations of the bait fisheries, it could not deny foreign fishermen access to the bait trade. Furthermore, a foreign nation retained the right to object to any legislation restricting such access on the grounds that the 1818 treaty between the United States and Great Britain created an international common property open to both parties. Any local legislation regarding the fisheries could not restrict American access any more than it restricted British access. Such an understanding could only be reached through a legal discourse on language. It is therefore essential that historians try to deconstruct this legal discourse in order to understand how the court could articulate such an interpretation for international management.

Both sides engaged in a lengthy debate concerning the language of the Convention of 1818. The key discussion throughout the arbitration focused on the word "liberty" versus the word "right." Determining what the negotiators of the Convention of 1818 intended when they penned the phrase "inhabitants of the United States shall have the *liberty*," and to what basis this phrase relied on the language of the 1783 treaty, which stated that "they shall forever have the *right*," became a principal focus of the debate. Similar debates also revolved around the words "bays," "coasts," "in common with," "grant," "inhabitants of," and just about any other significant word or phrase found in the Convention of 1818.

The British delegation argued that the term "liberty" in 1818 modified the term "right" in the 1783 treaty. The definition of "liberty" required a permission, whereas "right" was an open guarantee. The 1818 treaty, the British argued, restricted American access to a greater degree than had the 1783 treaty. Early in the opening of the British case, they argued that a permission to fish came with any regulation of the act that was permitted:

> Article one of the treaty under discussion gives to the inhabitants of the United States, liberty to fish on certain parts of the coasts of British territory. The term "liberty," as here used, is equivalent merely to permission. It is true that when granted by treaty it became as between Great Britain and the United States a matter of right, but there can be no question as to the extent of what was granted. It was merely permission to fish, in common

> with British fishermen, and was necessarily subject to the right of regulation by the Government of the country, inasmuch as, in the absence of such regulation, the subject-matter of the grant might itself be destroyed.[20]

The British jurists argued for a limited definition of the meaning of the word "liberty." Such a definition would of course be of great value for the British case because it would thus limit the rights of American fishermen along the treaty coast. The more the British lawyers could convince the court of a limited definition of the treaty as a whole, the more rights they and their Canadian and colonial counterparts would have over the government of the United States.

The American delegates sought to establish that the 1818 negotiators defined "liberty" in the same spirit as the 1783 negotiators had defined "right." This defined "right" as an absolute *right*, unquestionable and undeniable by any form of legislation. To defend their case, they focused on the "in common with" clause of Article 1. By linking the liberties granted to Americans with those "in common with" British fishermen, the American delegates argued that Americans had the right of legislative representation. Therefore, municipal legislation could not restrict American operation in the fishing zone beyond the actual act of extraction, because American fishermen did not have representation in that legislative body. They argued that the North Atlantic fishery, as stipulated in the Convention of 1818, constituted an international economic zone in which free enterprise of one group could not be restricted by the government of another group.

Under such logic, only the United States government held American fishermen accountable for their actions, and thus neither British, Canadian, nor colonial regulation could in any way restrict American fishermen, regardless of the jurisdiction of the territory under question. In his June 30, 1906, letter to Sir Edward Grey, the U.S. secretary of state Elihu Root argued that the 1818 treaty must be understood within the context of the definition of liberty, or as a right that cannot be restricted by unfair legislative representation. He argued that "The *liberty* to take fish shall be held in common, not that the *exercise* of that liberty by one people shall be the limit of the exercise of that liberty of the other." Root maintained that it was of no concern to American fishermen if or how Newfoundland fishermen exercised their share of the common liberty. Legislation regulating the conduct of Newfoundland bait fishermen did not in any way restrict the conduct of American bait buyers, each of whom held equal but separate rights to the fisheries: "Neither right can be increased nor diminished by the determination of the other nation that it will not exercise its right, or that it will exercise its rights under any particular limitations of time or manner."[21]

The British, however, maintained that "in common with" simply meant that the American fishermen would have to obey the same rules and regulations as British fishermen as set by the Canadian and/or British representative governments. In this sense, the British approached the issue of legislation in terms of their sovereign right. The idea that the inshore fisheries around Canada or Newfoundland constituted an international common property would violate British sovereignty in its territorial waters. The British jurists argued that international treaties did not supersede national legislation because "the mere grant of a right or liberty to subjects of one State to do certain acts in the territory of another State, does not itself confer any exemption from the jurisdiction of the State in which those acts are done." As such, the exercise of American rights under the granted privileges must be "exercised subject to such laws and regulations as apply to the subjects of the State which makes them."[22] In terms of international relations, the principles of common law did not override the importance of national sovereignty. Extending this argument to maritime jurisdiction, and to the specifics of the fisheries in question, the British delegates made a distinction between the deep-sea international fisheries and the national or domestic fisheries in inshore waters. Here again, we see these lawyers using words to divide the environments in the arena of international law.

The British lawyers argued that the 1818 treaty must be understood within the context of the 1783 treaty, which clearly made a distinction between the bank fisheries and the coast fisheries. For the bank fisheries, the 1783 treaty stated, "It is agreed that the people of the United States shall continue to enjoy unmolested the *right*," while for the coast fisheries the 1783 treaty more narrowly stated, "that the inhabitants of the United States shall have *liberty* to take fish of every kind." The British case claimed that "According to the contention of His Majesty's Government, the former was a recognition of a 'right' to fish in the ocean and the gulf: the latter was a liberty or license to take fish in British territorial waters conceded by Great Britain for political reasons."[23] Therefore, those coastal fisheries were still subject to British law without the necessity of consent from the United States government. While the Americans claimed that "liberty" meant "right," the British claimed that "liberty" meant "grant."

These lawyers constructed language to reconstruct an environment. While they debated such concepts as liberty and rights under law, none of them ever considered the nature of the fisheries. These lawyers proceeded to reconstruct the seascape according to legal phraseology. Throughout the nineteenth century, law and society attempted to shape the fisheries geography into terms they could not only comprehend but also manage, but never before had the ideas and ideals of law and legal precedent held such command over fishery management.[24]

In the cases and counter-cases, in the written and oral arguments, both American and British representatives referred to a legacy of legal phraseology dating back to the early seventeenth century in an attempt to show that their interpretation of the language of the 1818 treaty had clear legal precedent and proper historical context. The American case stated that the wording of the treaty must be understood in the context of the "well-known meaning which attached to them as used in legal phraseology."[25] Throughout all of these debates, the key to establishing the authority of each interpretation was to establish the historical context of the word or phrase itself. Thus, both sides submitted cases and arguments that referred to a variety of documents that each believed properly outlined the meaning of the words in the minds of the negotiators in 1818. For example, Charles P. Anderson, agent of the United States in the North Atlantic Coast Fisheries Arbitration, stated the following in his final report to the government of the United States:

> The determination of the true intent of the meaning of Article I of the Convention of 1818 with reference to the questions submitted for the decision of the tribunal, required an examination not only of the language of the treaty, but also of the events leading up to its negotiation and signature and of the actions taken by either Government since the date of the treaty having a bearing upon its interpretation.[26]

Thus, the history of language became a crucial element in the establishment of fishery management in the North Atlantic.

The central goal of the American delegation was to secure for the United States' fishing industry unrestricted access to inshore waters so as to freely participate in the all-important bait trade without influence from Britain, Canada, or Newfoundland. They claimed that access to those waters was a right rather than a grant. To accomplish this, they not only argued for their version of the definition of "right," but also maintained that the restriction of American rights to the inshore waters only came when the local governments or populations wanted to secure better commercial relations with the United States. This line of reasoning attempted to show that local economic bias structured the Canadian and Newfoundland legislation adopted to restrict American rights, and not the rationality and objectivity of the law as set by the Convention of 1818. By showing that the British and Canadian interpretation of the language did not have proper historical context or legal precedent, the American case could establish that a precise reading of the law did not justify its opponent's case.

The Americans illustrated their argument by showing that no local government placed any restrictions upon American fishermen within British territorial waters for the first twenty years after the Convention of 1818. Furthermore, the Americans argued, the British Act of June 14, 1819—an act that gave the Convention of 1818 legal power in British North America—was intended not to regulate American fishermen within the territorial limits of Britain, but to restrict British fishermen from interfering with American rights. According to the American case, the clearly accepted initial interpretation of the treaty by the British authority stated that the local governments held no legal rights or authorization to restrict in any way American access to the fisheries that the terms of the treaty gave them. The Americans laid out the importance of the "historical" context of the 1818 agreement by noting that "no attempt was made prior to 1836 to establish by provincial legislation any regulations or restrictions upon American fishermen resorting for the four purposes mentioned in the treaty to the bays and harbors referred to in the last part of the fisheries article of the treaty."[27] Only after 1836 did local legislation seek to limit American rights as set by the 1818 agreement. Therefore, local legislation was not structured according to an objective interpretation of the law.

The 1836 provincial legislation of Nova Scotia intended to restrict American access to British territorial waters. This legislation, the Americans argued, gave a new and unauthorized interpretation to the language of the Convention of 1818. The Nova Scotia House of Assembly designed the legislation to give an unambiguous advantage to local British fishermen over American fishermen within the North Atlantic fisheries, which the 1818 treaty intended to keep open to both parties as a free and common resource.

> The desire of driving American fishermen away from their coasts altogether, and the opportunity of securing half the profits from seizures under the act of 1836 offered every inducement to the people of Nova Scotia for searching

> the treaty for some new meaning favorable to their purpose. . . . This new interpretation was not officially communicated to the United States by Great Britain until several years later, but rumors of its purpose and effect soon reached the United States.[28]

By adopting a new interpretation of the language of the international agreement, the Nova Scotia legislation, according to the American case, violated not only the wording of the treaty but also the intent and spirit of the mutual convention.

The Americans argued that as a "reinterpretation," the Nova Scotia case was unjustifiable; furthermore, Nova Scotia had no right to construct any interpretation of the treaty, because that government body was not an original contracting party. Nova Scotia, the Americans argued, had no right to interpret or police an agreement signed between the United States and Great Britain. Their policing efforts were not only unauthorized by the terms of the treaty, but the aggressive spirit in which they went about this enforcement constituted an unfriendly and uncivilized act. An American jurist argued the following before The Hague in 1910:

> Not content, however, with their efforts to prevent the American fishermen from exercising the treaty right referred to, the provincial authorities proceeded to devise a new and far reaching interpretation of the renunciatory clause of the treaty in order if possible to deprive the American fishermen of their right in the outer portion of the great bays or indentations of the Nova Scotia coast, in which the American fishermen had always claimed and exercised the right of fishing beyond the limit of three miles from the shore.[29]

Illustrating that local bias constituted the justification for any restrictive interpretation of the Convention of 1818 greatly aided in their overall case.

The American jurists at The Hague introduced an important case of legal precedent by their citation of the decision of the Vice-Admiralty Court in Halifax in 1853. In that year, the Claims Commission of that court refused to recognize the seizures of the American schooners the *Washington* and the *Argus*. The court's umpire, Joshua Bates, awarded the owners of the *Washington* $3,000 and the owners of the *Argus* $2,000 for damages from the unlawful seizures, noting that the Convention of 1818 did not authorize Nova Scotian authorities to seize American vessels. The policy of the British Admiralty throughout the 1870s, furthermore, reinforced this interpretation when the British marine authority emphasized a more relaxed enforcement of their territorial fisheries while the Canadian Marine Police pursued a more aggressive stance. By exposing the contrast between the two policies, the Americans claimed that the more limited interpretation of American rights within territorial waters clearly resulted from local or regional bias and not from an objective reading of the law. Furthermore, the Americans argued, since neither the local nor the national governing bodies in North America held any legal jurisdiction in the matter of international agreements, the interpretation of American rights by the British Admiralty represented a more accurate reading of legal precedent.[30]

The American arbitrators continued to emphasize that the more stringent interpretation of the language used in the Convention of 1818 did not reflect a proper legal discourse. Instead it resulted from intense local bias and regional objectives to control the entirety of the fisheries and secure its profitable returns. This intent clearly violated the "in common with" clause of Article 1 of the Convention of 1818, the Americans argued, because the framers designed that clause to provide for full and equal access to the fisheries for both British and American fishermen. The case of the United States continued to push its interpretation as it cited examples of local interference of American fishing efforts. In reference to the Fortune Bay incident in Newfoundland, the American arbitrators argued that "The action of the Newfoundland fishermen in preventing the American fishermen from taking advantage of their long awaited opportunity could hardly have been inspired by a respect for law and order." Throughout the case, the Americans argued for the removal of managerial and policing power from the hands of local governments or local populations that could not properly organize the extraction of resources along what the American negotiators believed to be fair and honest lines.[31]

In addition, the Americans also argued that the bias of economic interest propelled the Canadian government's refusal to recognize that American fishing schooners retained the commercial rights that all other American vessels had within the ports of British North America, and later Canada, as a result of the commercial treaty signed between the United States and Great Britain in 1830. The American position maintained that these free port acts extended commercial rights to all American vessels carrying the proper licenses as issued by the United States government, whether they be fishing schooners or merchant ships. Under these trading rights, American fishing schooners used the local ports and harbors in Newfoundland and the Maritime Provinces as bases of operations from which they secured bait, ice, and other supplies. In addition, these schooners transshipped their fish cargoes to steam vessels or railroads for shipment to the United States. The Canadian authorities argued that this trafficking violated the Convention of 1818, which clearly stated that American fishing vessels could utilize British harbors for procuring wood and water, for repairing damages, or for shelter from storm, and for no other purpose whatsoever. The Americans argued that the trade treaty of 1830 amended the Convention of 1818 and allowed for free and open ports for fishing vessels in the North Atlantic. Canada's interpretation of the 1818 agreement, the Americans argued, derived from their desire to force the United States government into a new commercial arrangement and was therefore not the result of an objective reading of the language. Thus, the closure of the ports to American fishermen, and the general hostility towards American fishermen, reflected the bias of Canada's fishery policy and a violation of both the intent and the language of the Convention of 1818.[32]

The American case argued that when it became clear that a new commercial treaty with the United States was not possible, the Canadian government no longer enforced its interpretation of the 1818 treaty. By adopting the modus vivendi of 1888, the Dominion recognized the futility of their interpretation. By opening their territorial waters to American fishermen without harassment from local marine police, Canada essentially agreed with the American interpretation of the Convention of 1818. Trouble between American fishermen and Dominion marine police ended in 1888. The American arbitrators claimed that the court must embrace their argument because the 1888 modus vivendi already recognized the American interpretation. Moreover, as long as the North Atlantic fisheries remained free and open to American use, as outlined by the Convention of 1818, there would be no controversy between Canada, Great Britain, or the United States.

> So far as Canada is concerned this modus vivendi has been continued in practical effect down to the present time by action of the Canadian Government without formal extension, and during this period no change has taken place in the attitude either of the United States or British Government on the questions of difference which had previously arisen in the fisheries controversy with reference either to the so-called treaty coasts or other coasts of Canada.[33]

The 1888 modus vivendi illustrated the Canadian acceptance of the American interpretation, and it signaled that Canada only sought a more stringent reading of the law in the years 1836–1854, 1866–1871, and 1885–1888, or whenever Canada sought more favorable trading relations with the United States. Thus, the Americans argued, Canadians' and Newfoundlanders' economic bias undermined their reading of the language of the 1818 treaty.

In Newfoundland, however, the local government remained hostile to the presence of American fishermen within their local waters throughout the entire period. The American arbitrators at The Hague needed to show, as they had with Canada, that local bias was the basis for the exclusionary legislation adopted by Premier Bond in Newfoundland, that this legislation had no historical or legal precedent, and that it therefore did not represent an objective reading of the language or the intent of the Convention of 1818. The 1905 Newfoundland legislation that restricted American rights in Newfoundland waters, the Americans argued, was the result of the intent of that government to secure more favorable trade relations with the United States. The fact that it came on the heels of two failed commercial treaties between Newfoundland and the United States provided strong support for the American case at The Hague, which argued that the 1905 legislation was a harassment policy specifically adopted by Newfoundland "to compel the American Government to open the American markets to Newfoundland fish and fish products free of duty in exchange for more extensive fishing and commercial privileges; and it soon became evident that in furtherance of this plan the Newfoundland Government was likely to attempt to impose certain limitations and restraints upon the American fishermen in the enjoyment of the fishing liberties secured to them under the Convention of 1818 upon the treaty coasts of Newfoundland."[34] All local legislation, the Americans argued, had to be constructed to avoid any violation of the terms or meanings of the Convention of 1818. But this 1905 law in Newfoundland not only violated the language of the treaty, they reasoned, it also undermined the intent of that treaty to create an international and common resource.

The international agreement superseded any local legislation. The American case claimed that the 1818 treaty became the permanent legislative authority for the international common property in the North Atlantic. As such, the rights granted to the United States by this agreement were beyond the sovereign power of Great Britain to change or adapt in any way. As the Americans argued, this amended British territorial rights:

> It is conceded that this right is, and forever must be, superior to any inconsistent exercise of sovereignty within that territory. The existence of this right is a qualification of British sovereignty within that territory. The limits of the right are not to be tested by referring to the general jurisdictional powers of Great Britain in that territory, but the limits of those powers are to be tested by reference to the right as defined in the instrument created or declaring it.[35]

The United States always argued that the "in common with" clause of the Convention of 1818 meant that American and British fishermen should enjoy complete equality in their rights to the fisheries and in the management of that resource regardless of territorial authorities. Any legislation in Canada or Newfoundland would obviously be adopted for the benefit of the local population. Such legislation had only to gain the consent of the constituents, and since the American fishing laborers had no representation in Canada or Newfoundland, those legislative bodies could not adopt policies that restricted American fishermen.

The American case rested on the interpretation that within the international common property of the North Atlantic fisheries, only the American government retained the legal right to restrict the actions of American fishermen. Therefore, if anyone desired to manage the fisheries, that management must come not from local legislation, which carried the danger of local or regional bias, but from the mutual consent of the contracting parties. In 1906, when Elihu Root was secretary of state, he expressed this American belief when he wrote Sir Edward Grey in London, stating:

> For the claim now asserted that the Colony of Newfoundland is entitled at will to regulate the exercise of the American Treaty right is equivalent to a claim of power to completely destroy that right. This Government is far from desiring that the Newfoundland fisheries shall go unregulated. It is willing and ready now, as it has always been, to join with the Government of Great Britain in agreeing upon all reasonable and suitable regulations for the due control of the fishermen of both countries in the exercise of their rights, but this Government cannot permit the exercise of these rights to be subject to the will of the Colony of Newfoundland.[36]

Root and others argued that the Americans had no desire to open the fishery resources to unregulated extraction, only that any regulation must be done in cooperation with other relevant powers.

In opposition, the British case argued that the American "right" constituted merely a "grant" to access the fisheries within British territorial waters, and therefore these fisheries were not free fisheries but regulated by British authority. British, Canadian, and colonial authorities, therefore, retained the right to legislate their own resources in order to ensure that those resources would be protected and preserved. The British interpretation of the "in common with" clause stated that American fishermen must obey the same legislative restrictions as British fishermen. As with the American case, the British jurists cited legal precedent to successfully defend their interpretation.

The British jurists introduced into evidence several diplomatic exchanges between the United States and Great Britain to prove that the issue of local jurisdiction did not become a matter of international controversy until American fishermen encroached upon local waters. Only after this violation of local property did local legislative bodies seek to rebuke American "aggression." The legislation did not respond to the national goals of increased commercialization, as the Americans had argued, but addressed foreign aggression that violated the terms of the treaty. United States authorities presented no objection to local legislation until the domestic American fisheries resources collapsed in the 1870s, thereby forcing American fishermen into foreign fisheries. Therefore, the British argued, the Americans did not base their interpretation of free access to local fisheries in British North America on a proper reading of the law, but instead upon the need of American industry to find new fishing grounds. This interpretation, the British jurists argued, did not develop as the official policy of the United States until it was communicated via diplomatic channels in 1878. Thus from 1818 to 1878, the first sixty years of the Convention of 1818, American governing officials accepted the British interpretation of regulated fisheries. The British arbitrators argued that "In 1856, and on several occasions between 1866 and 1872, the attention of the United States Government was called to the necessity of fishermen obeying Colonial laws, but no objection was taken. It was not until 1878 that the present contention was first raised."[37] Thus the British asserted that the American interpretation as presented at The Hague in 1910 did not represent the original accepted interpretation of the Convention of 1818.

Throughout the entire historical period of the Convention of 1818 (1818–1854, 1869–1871, and 1885–1910), the British argued, the official policy of Great Britain maintained that the North Atlantic fisheries within their territorial jurisdiction was a regulated fishery, and not free and open to unrestricted use by American fishermen. Although the governments of Great Britain, Canada, and the colonies may have from time to time considered it reasonable to relax certain regulations or to enforce more stringent codes of extraction, the British negotiators maintained, the fishery remained regulated by law. Nothing in the historical documents, the British argued, undermined the claim that Great Britain always viewed their fisheries as a regulated resource.

The British based this policy upon the belief that the Convention of 1818 could only be understood in the context of the 1783 treaty. The Treaty of 1783 allowed American access because, as former British subjects, New Englanders had won the exclusive use of those fisheries from the French. Moreover, as former British subjects, they retained the right to the use of the common water resources. If their former status as British subjects provided the basis for their right to the common resources, then that same condition forced them to obey certain laws and statutes as established by the British government. Sir Edward Grey, secretary of state for foreign affairs in 1906, best communicated this understanding in a letter to Elihu Root:

> American fishermen cannot therefore rightly claim to exercise their right of fishery under the Convention of 1818 on a footing of greater freedom than if they had never ceased to be British subjects. Nor consistently with the terms of the Convention can they claim to exercise it on a footing of greater freedom than the British subjects "in common with" whom they exercise it under the Convention. In other words, the American fishery under the Convention is not a free but a regulated fishery, and, in the opinion of His Majesty's Government, American fishermen are bound to comply with all Colonial Laws and Regulations.[38]

The British emphasized that the "in common with" clause of the Convention of 1818 had always been the key modifier of the entire treaty. This one clause, the British argued, restricted American access to the fisheries along the same lines as British subjects. Any law regulating British subjects' rights to extract resources from the fisheries also regulated American fishermen.

Not until the 1870s did the United States government refuse to recognize this right. For British jurists, this late date clearly proved that the American policy did not properly rest on the language of the treaty itself. All other negotiations regarding the fisheries—the Reciprocity Treaty of 1854, the Treaty of Washington of 1871, and the failed treaties of 1888, 1900, and 1905—maintained the British view of the "in common with" clause, that "American fishermen pursuing their occupation within British territory would be bound to observe the local laws and regulations in like manner as all foreigners are bound to observe the municipal laws of the country in which they are resident."[39] Thus, the British countered the American claim by stating that all foreigners must obey certain domestic laws while they visit foreign nations. This dynamic remained especially true when those foreigners extracted wealth from the resources of other countries.

The British emphasized that no historical document contained any proof that Great Britain resigned its right to regulate commerce and fishing within its own territories. This right to regulate the fishery industry, the British observed, was clearly laid out in the terms of the Convention of 1818 and all subsequent communications by British officials. The British maintained this right to regulate for the sole purpose of preserving and protecting the resource-based economy, and not, as the Americans had argued, to give any unfair advantage to British subjects at the expense of American fishermen who worked in the common waters of the North Atlantic. In this argument, the British jurists believed that "Great Britain does not claim the right to destroy that which is conceded by the treaty to American fishermen, but the right only to make regulations which are necessary or desirable for the preservation of the fisheries for the benefit of British and American fishermen alike."[40] Thus, the British arbitrators did not reject the idea that the fisheries under question were in fact common resources and open to American use. Instead, they maintained that the British government retained certain rights to regulate these common resources.

While the British jurists conceded the idea of a common property within their own territorial waters that allowed American fishermen certain rights, those rights had to be limited according to British law. Americans were not, under the Convention of 1818 or any other diplomatic agreement, exempt from the law of Great Britain. The British jurists concluded that "there is no suggestion of any exemption from British jurisdiction, nor is there any expression in that article, or any other article, which in any way points to such a construction."[41]

The debate continued along this course for weeks. The Americans continued to argue that the liberty granted to them by the Convention of 1818 was a right of free access to a common property that could only be limited by a legislative body that represented them as equal to all other involved parties. Since the American fishing industry had no representation in British, Canadian, or colonial legislative bodies, then those bodies had no right to restrict the Americans' rights of access. Any laws constructed by those governing bodies regulating the international bait trade, the American lawyers maintained, had been and would continue to be constructed with a clear bias for local interests. The free rights of Americans were thus restricted. This, therefore, would violate the terms of the Convention of 1818, which prescribed Americans' rights as "in common" with those of British fishermen. The American arbitrators defended their interpretation of the Convention of 1818 by showing that national economic or political goals negatively affected the opposition's analysis, which therefore was not based on an impartial and logical reading of the precise language of the treaty itself.

The British agreed that the Americans had the right to access the common property granted to them by the Convention of 1818, but they also maintained that the British government could restrict that right within the same parameters they felt necessary to govern their own subjects. American fishermen did not hold any additional rights to the fisheries than those granted to British subjects. Such a case would give the Americans an unfair advantage. As the fisheries in question remained within the jurisdictional authority of not only Great Britain but also the colonial government of Newfoundland and the government of Canada, those legislative bodies retained the right to regulate the fishing business of all fishermen within those waters whether they be local subjects or foreign citizens. Such was the policy of Britain from the beginning of the protracted debate on the North Atlantic fisheries. The British case argued that based on a judicious reading of the terms of the Convention of 1818, the United States fully accepted that policy until circumstances beyond the language of the treaty itself led to the Americans' attempt to secure more rights than could be allowed by an objective reading of the 1818 treaty.

Both sides argued for the creation of a management policy based solely on their interpretation of what was meant by the language of the Convention of 1818. This focus resulted in the creation of a legal interpretation that reinforced the idea of an international open-access resource; it defined a resource that allowed national legislation to retain certain controls over extraction, but only if those controls did not restrict American free access to the trade based on that extraction. The Hague decision in 1910 that allowed the United States to appeal Canadian or provincial legislation to an international joint commission set the fisheries of the North Atlantic clearly within an international context that continued to allow nearly unlimited access to those fisheries by the numerous competing factions—both inshore and offshore, American and Canadian—for the next sixty-seven years. The Hague decision in 1910 thus opened the resources of the great oceans to an unfettered amount of extraction throughout most of the twentieth century. Any attempt to limit access to the international common fisheries would plague international fishery commissions, agencies, arbitrations, and committees for generations until 1977, when the United States and Canada set strict terms of access to all fisheries within two hundred miles of the coast. Significantly, both were later forced to uphold their decision with military force.[42]

Conclusion

After the 1910 decision at The Hague, the bait trade between American and Canadian fishermen continued in its usual manner. Americans were permitted to enter local waters to purchase bait and supplies so long as they did not interfere with the exclusive rights of local fishermen to harvest the catch themselves. After a hundred years of diplomatic communications, three treaties, three failed treaties, and a series of modi vivendi, the actual operations of the bait trade had changed very little. The reason for this is because local fishermen succeeded, by about 1830, in creating and enforcing their own informal codes of conduct. These rules limited extraction within the local environment to the local population, while at the same time allowing that population to participate in the global fish trade. Local bait fishermen and their international customers successfully navigated both national politics and international diplomacy, as well as the various legal codes established by both, to create a unique economy based on limited resource extraction and community-defined stewardship.

Those communities involved in the North Atlantic fisheries created locally defined informal codes of conduct—rules created through collective action, peer pressure, crowd protest, or intimidation that were outside or beyond the formal laws or regulations as set by local, national, or imperial legislation, or by international treaty. In 1836 mackerel fishermen on the south shore of Nova Scotia sought to enforce these informal codes of conduct when they refused to sell their catch to Philip Carten and instead decided to wait for the arrival of the American fishing fleet. In 1871 merchants in Charlottetown, Prince Edward Island, enforced these informal codes of conduct when they transshipped American-caught fish in local ports regardless of the protests filed by the officers of the Royal Navy. In 1877 local weir fishermen in Fortune Bay, Newfoundland, enforced these informal codes of conduct when they prohibited American fishermen from catching fish in their own immediate environment, even though those Americans held treaty rights that permitted them to fish. In 1886 fishermen and farmers in and around Digby, Nova Scotia, enforced these informal codes of conduct when they decided to sell baitfish and ice supplies to the *David J. Adams* despite local and national legislation that defined such acts of trade as smuggling.

All of these efforts emerged within the international context of what became known as the "fishery question." Although those specific words, "fishery question," did not appear regularly until the heated years of 1885–1886, the debate was well in place by the 1830s and would last until 1910, when it was settled by the arbitration hearing at The Hague. The question was: what rights or privileges do American fishermen have in the inshore waters and ports of British North American and Canada? Although the Convention of 1818 outlawed Yankee efforts to fish the inshore waters, a heated debate about their right to use local ports as bases of operations in which they would hire crew members, trade with local communities, and purchase supplies—including ice, and most importantly baitfish—continued throughout the period. Due to international treaty stipulations that effectively formalized earlier informal codes of conduct, the debate subsided briefly between 1854 and 1866, and again between 1871 and 1885. In Canada, the diplomatic question came to a conclusion after the adoption by the Dominion government of the modus vivendi of 1886 and its continuation well into the twentieth century. In Newfoundland, the debate remerged during the first decade of the twentieth century, finally coming to some conclusion in 1910. Because the North Atlantic fisheries played a vital role in shaping Anglo-American relations throughout the nineteenth century, the fishery question received extensive documentation in the press, in political debates, in legal records, and in diplomatic correspondence.

When we examine the documents related to the political and diplomatic history of the North Atlantic fisheries, we uncover a history far more complex than it appeared on the surface. Laborers, merchants, and farmers all played roles in shaping international relations in the border seas of the North Atlantic. Although diplomats created the Convention of 1818, the Reciprocity Treaty of 1854, the Treaty of Washington of 1871, the modus vivendi of 1886, and The Hague arbitration decision of 1910, as well as numerous politically created pieces of legislation, the actual business of fishing and selling fishing commodities was largely shaped by the working people in the communities that bordered the North Atlantic fishing grounds. By looking deeper into the documentation regarding policy, diplomacy, and law, we can uncover the labor history of resource extraction within this diplomatically charged environment.

These laborers structured the fisheries along two consistent policies. The first limited actual extraction of fish from the water to the local population. This was a unique policy related exclusively to the inshore fisheries, and particularly to the bait fisheries. Until the end of the twentieth century, fishing for the great codfish remained an international affair largely outside the parameters of individual state action. Grand Bank fishing attracted a host of operators throughout the twentieth century that included the British, Canadians, Americans, Spanish, and French—a list that greatly expanded in the twentieth century, especially after the end of World War II. Throughout all of this, the inshore fisheries of British North America and Canada remained under the exclusive jurisdiction of Great Britain and/or Canada. Yet, Americans were always eager to break into this geography. The inshore fisheries possessed a rich resource of small schooling fish like herring and mackerel that migrated inshore for spawning. Local communities sought to retain control over these fisheries regardless of the political or diplomatic maneuvers of the nation-state leaders. When the Reciprocity Treaty of 1854 and the Treaty of Washington of 1871 granted to the American fishermen rights to access these waters, local fishermen often responded through community action, intimidation, or outright violence to reinforce their own local codes of conduct that maintained the inshore fisheries for the exclusive use of the local community.

Even more important than the rights or limits of outsiders to fish in inshore waters were the rights or limits of those outsiders to come into the inshore waters for other purposes related to the fisheries. In these ports, the Americans could transship catches back to home markets, hire crew members, and purchase supplies such as bait and ice. Such access was essential to the American fishing efforts. When national and local governments sought to enforce a strict interpretation of the Convention of 1818 by prohibiting American fishermen from entering local waters or ports for any purpose other than wood, water, shelter or repair, Americans found numerous local fishermen eager to cooperate with the Yankees in their efforts to deceive the local law-enforcement efforts. The reason for this was not, as many politicians claimed, because the local fishermen were unpatriotic or otherwise blinded by greed. These local fishermen only sought the continuation of their established informal codes of conduct. Such rules allowed outsiders to enter local waters for trade so long as those outsiders did not attempt to actually fish in the local waters. This constituted the second overarching informal code of conduct throughout the nineteenth century.

The combined effect of these two informal codes of conduct, the restriction of fishing to local fishermen and the permitting of trade to outsiders, operated outside the limited boundaries of formal law and held important regulatory power throughout the nineteenth century. Despite enormous political and diplomatic efforts to structure law and order in the fisheries around nationality, the actual business of fishing and fishing-related commerce remained largely defined according to informal codes of conduct that only loosely, if at all, related to nationality. Local fishermen protested the fishing activities of all outside fishermen, foreign and domestic alike, while retaining the right to trade with foreign buyers despite national or imperial agendas to restrict such trade to domestic merchants. This practice began in the early nineteenth century as American fishing schooners arrived in the inshore waters of Nova Scotia looking for both fish and fishermen. After Nova Scotia fishermen took employment aboard the Yankee vessels, they were better able to establish trade relations with the local communities. This cross-border migration of labor thus directly contributed to the increase in commercial relations between New England deep-sea fishermen and Atlantic Canadian inshore fishermen and their home communities. The commercial relationship developed throughout the middle decades of the nineteenth century until it became such a regular occurrence that it was nearly impossible to prohibit with legislation, law, or police power. By the beginning of the twentieth century, international law could only formalize the preexisting informal codes.

The resistance to the "intrusion of strangers" so common among local fishing communities in the northwest Atlantic also had a decidedly anti-industrial element to it. Towards the end of the period, new technologies were introduced so as to increase the catch per unit of effort in the fishing business. The process is well known to most maritime and environmental historians, but it has not been fully incorporated into a more general history of labor and international relations in the North Atlantic. The resistance to the introduction of technologies such as the purse-seine net of the steam-powered bottom trawler among local fishermen in Canada, as elsewhere, has often been interpreted as a case of moral economy in which pre-industrial laborers resist the development of industrial capitalism and its exploitative tools.[1] This, to a large extent, is an accurate interpretation of the trend. Yet, a more complete review would have to include earlier efforts among local fishing communities to retain control of their resources in the face of opposition from not only foreign capitalist engines, but also their own national or state governments and their international representatives, who often sought treaty negotiations that would open the resources that the local communities strived to keep closed.

The steam-powered bottom trawler is a fine example of a new technology that had the potential of revolutionizing the modes of production of small-scale, localized fishing operations.[2] In 1910 the trawler *Wren* began operations out of Halifax, Nova Scotia. In that year, the Canadian Department of Marine and Fisheries reported the protest of "shore fishermen" who argued "that large quantities of small unmarketable fish were taken and wasted, and that she was destructive to the gear of the hand liners, as well as of the grounds resorted to by them for fishing purposes."[3] The Canadian government had banned the operation of the trawl within three miles of the coast, but the owners petitioned the government to allow for such operations only during the winter months, when the open seas were too rough for effective trawling. The owners argued that the trawl did not affect the inshore fisheries because it sought to catch sole and plaice, fish not marketed by the inshore fishermen. The owners argued that since the trawler and the inshore fishermen operated in different fishery industries, there should be no protest to their free use of the fishing grounds.

Despite these protests, the Dominion government sought to protect the fishing of the inshore fishermen within the three-mile limit. No effort, however, was made by the Canadian government, or by Great Britain in its own waters across the Atlantic, to limit the efforts of the trawlers beyond the three-mile limit. The government argued that without international cooperation, there would be little gained from the restriction of domestic trawling while foreign trawling went on outside Canada's territorial limits. French trawlers were busy at work throughout the North Atlantic from their western bases of St. Pierre and Miquelon off the southwest coast of Newfoundland. The issue, according to the Department of Marine and Fisheries, "was complicated by its international character." As such, the department hoped to utilize the newly created International Fisheries Commission to address the matter.[4]

Certain officers within the department—including William Wakeham, who championed the protection of inshore fishermen throughout the 1890s—saw no conflict with the introduction of the steam trawler. In 1910 he argued that the trawl could never operate in inshore waters due to the rough and uneven bottoms, and that the fish caught by the trawl were those not marketed by the inshore fishermen. Therefore, Wakeham concluded that no restrictions on the trawlers should be imposed, and they should have free range over the ground fishery at sea, where they caught profitable yet unmarketed flatfish like flounder. Wakeham suggested that Canada follow the example of Great Britain, which had already declared the trawlers to be sound engines for sustainable fisheries and were allowing over a thousand trawlers to operate in the North Sea.[5]

The Canadian government did follow the example of Great Britain by launching its own investigation into the effects of the trawler. In 1910 John J. Cowie submitted his report on the beam and otter trawlers. Cowie addressed "the cry . . . raised . . . that with the increased operations of steam trawlers, the sea will become depleted of fish, and that line fishermen will awake some day to find their occupation gone" in much the same way Britain had done. Although Bowie supported the exclusion of the trawler from inshore waters, he "cannot agree with those who assert that trawling, if long continued offshore, as well as inshore, is destined to deplete the sea of food fishes." Cowie further noted that the natural climate of the Grand Banks and other offshore fisheries of the northwest Atlantic effectively prohibited any fishing from December to May and so effectively closed the fishery for temporary relief that no formal legislation of restriction would be necessary. In conclusion, Cowie succinctly stated, "It is therefore inconceivable that trawling can develop to such a degree, on this side of the Atlantic, as to ever appreciably diminish the extraordinary abundance of certain classes of fish such as cod and haddock, in the waters of Canada, and I think we may rest assured that its fisheries will remain a splendid heritage for all time."[6]

From this brief review of the introduction of the trawler, which itself deserves a much more comprehensive history, it is clear that small-scale local fishermen in Canada had successfully gained political control over the inshore fisheries. Despite the economic pressure, as well as the national agenda of Ottawa, to increase Canada's fisheries production and to fully take command of its own domestic market, the highly effective trawler was banned from the inshore waters. This was certainly partially the result of the perceived inability of the trawler to operate effectively in these waters. Yet, the trawler owners were eager to be permitted into those waters, for at least the winter months, and thus naturally applied whatever political pressure they could to achieve that goal. This pressure, however, was not effective enough to outweigh the historical control of the inshore waters by local fishermen and their informal codes of conduct, which by 1900 had become codified in national and local legislation. Yet the effectiveness of these codes of conduct to facilitate the more general sense of environmental stewardship over fisheries resources began to wane. Although the inshore fishermen of Canada had clearly gained some control over fisheries in their local environment, it became increasingly clear after 1910 that such control was of only limited power, as the fisheries beyond the local environment became increasingly exploitative and would eventually impact their local resources as well.

In addition to the use of the purse-seine net and the steam-powered trawler, the rapid rise of canning came to challenge the local interpretation of the baitfish supplies. With canned sardines came a new use for juvenile herring. Some historians have begun to examine the complex history of the sardine business.[7] Towards the end of the nineteenth century, numerous factories appeared along the coast of northern New England and Atlantic Canada as part of what became known as the "Portland Syndicate." These factories, both in the United States and Atlantic Canada, were under the direction of a handful of financial elites based out of Portland, Maine. To a large extent, this trust controlled both wholesale and retail pricing. They thus extended their control over the environment itself by, to a certain extent, dictating production. Weir fishermen in both countries launched a series of petitions and on occasion used social intimidation or violence to regain control of their local environment. This effort largely failed in the face of the growing might of industrial capitalism in the late nineteenth century. This history deserves its own complete study. It is suggested, however, that any such study would have to emerge within the historical context of the fight for local control of local resources through the creation and enforcement of informal codes of conduct, which began as early as 1830.

In many ways, the present work has been theoretical. It presents a large, generalized idea concerning environmental stewardship among working people. It builds upon the existing scholarship while extending that work into new environments and geographies. It combines environmental history, maritime history, diplomatic history, and labor history into one story about environmental territorialism. Its purpose is to tell the story of the baitfish trade while also introducing this concept of environmental territorialism in the hope that others may find it appropriate in their own studies. Territorialism is defined as the extension of state power over a particular property. Environmental territorialism is the application of non-state power structures, defined by an interpretation of environmental stewardship beyond nation-state law, over that disputed property. Local fishermen in the northwest Atlantic sought to impose their own power system to control the property in an effort to seek long-term control and stability, something we may accurately define as a system of stewardship, even if the historical actors never used such terminology. There were certainly economic motivations in this effort. Fishermen were not environmentalists as we may define that word today. This, however, does not diminish their stewardship goals. We should not impose modern environmental standards to assess the efforts of past resource users. Local fishermen voiced their concern for resource stability through petitions, intimidation, and social violence, all in an effort to stabilize their economic base, which happened to be baitfish. These fishermen were not anticapitalist. The evidence clearly shows their eagerness to participate in the global fishery economy as it was structured during the nineteenth century. They actively sought out new employment, markets, and cash in exchange for their services. They did, however, seek to retain control over the tools of their trade by limiting access to the resources to local fishermen only, and by resisting the introduction of new modes of production like the purse-seine net or the steam-powered trawlers. There is a sense of a "moral economy" here, but they were not Luddites or antimodernists attempting to deconstruct the creation of a capitalist order. They only sought to retain some individual control within that capitalist system and thus moderate the effects of an economy of exploitation. They sought local control of resource production while at the same time seeking out global markets. Their success in doing so for as long as they did makes this a unique history.

In the end, this history hopes to warn environmental historians not to discount the efforts of working people. Individuals who catch fish, cut down trees, or reap plants do not necessarily view themselves as laborers in an industrial machine. Nature was not necessarily viewed as just a place to extract wealth and nothing more. The fishermen of this history clearly understood their local waters to be something more than a common resource to be exploited quickly in order to extract as much individual wealth as possible before someone else got to it.

Their efforts to limit both access and production were a clear, if not conscious, effort to address the tragedy of common property management.

This work also seeks to provide a counterbalance to the growing faith in international or global history. Certainly global awareness is important, but we cannot forget that many people lived in a much more regionally defined world. Fishermen held a local perception of the environment that went beyond simplistic economics to include much more complex cultural constructs. There was a sense of community identity among the inshore fishermen, an effort to preserve a resource for a community and for the future generations within that community. By gaining environmental territorialism over their immediate waters, they retained a sense of allegiance to that environment. Perhaps it was a sense of ownership, but this term is too loaded with modern capitalist meaning to be used to define the cultural construct. Instead, environmental territorialism seems to work best. These fishermen created systems of control over the property based on their interpretation of the meaning of the environment and a perceived need to seek out successful methods of stewardship. The eventual failure to retain control should not demean their efforts. We can learn much from the pride they had in their work, and their appreciation for the waters lapping up to their front doors, and their recognition of the value and importance of the fish in those waters.

Notes

INTRODUCTION

1. Sean Cadigan, *Hope and Deception in Conception Bay: Merchant-Settler Relations in Newfoundland, 1785–1855* (Toronto: University of Toronto Press, 1995); Stephen J. Hornsby, *British Atlantic, American Frontier: Space of Power in Early Modern British America* (Hanover, NH: University of New England Press, 2005); Harold Innis, *The Cod Fisheries: The History of an International Economy* (New Haven, CT: Yale University Press, 1940); Mark Kurlansky, *Cod: Biography of the Fish That Changed the World* (New York: Penguin Press, 1997); Wayne O'Leary, *Maine Sea Fisheries: The Rise and Fall of a Native Industry* (Boston: Northeastern University, 1996); Rosemary Ommer, *From Outpost to Outpost: A Structural Analysis of the Jersey-Gaspé Cod Fishery, 1767–1886* (Montreal: McGill-Queen's University Press, 1991); Peter Pope, *Fish into Wine: The Newfoundland Plantation in the Seventeenth Century* (Chapel Hill: University of North Carolina Press, 2004); Daniel Vickers, *Farmers and Fishermen: Two Centuries of Work in Essex County, Massachusetts, 1630–1850* (Chapel Hill: University of North Carolina Press, 1994).
2. George Brown Goode, *The Fisheries and Fishing Industry of the United States*, vol. 5, *History and Methods of the Fisheries* (Washington, DC: Government Printing Office, 1887), 151.
3. In 1910 the Canadian Department of Marine and Fisheries reported that in Canada, inshore fishermen outnumbered deep-sea fishermen by eight to one. "Forty-Second Annual Report of the Department of Marine and Fisheries, 1908–9, Fisheries," *Sessional Papers*, 1910, no. 22, xxix.
4. Goode, *The Fisheries and Fishing Industry of the United States*, vol. 5, *History and Methods of the Fisheries*, 152.

5. James Acheson, *The Lobster Gangs of Maine* (Hanover, NH: University Press of New England, 1988); Acheson, *Capturing the Commons: Devising Institutions to Manage the Maine Lobster Industry* (Hanover, NH: University Press of New England, 2003).
6. "Article I, Convention of 20 October 1818, between the United States of America and His Britannic Majesty," in Joseph Pope, *A Short Account of the Negotiations Which Resulted in the Convention of the 20th October, 1818, between Great Britain and the United States, Together with the Text of the Treaty* (Ottawa, 1907).
7. Garth Stevenson, "Canadian Regionalism in Continental Perspective," *Journal of Canadian Studies* 15, no. 2 (Summer 1980): 17.
8. C. Vann Woodward, *The Comparative Approach to American History* (New York: Basic Books, 1968), 3.
9. Stevenson, "Canadian Regionalism in Continental Perspective," 18.
10. Victor Konrad, *Borderlands: Essays in Canadian-American Relations* (Toronto: E.C.W. Press, 1991), viii.
11. Donald Worster, *Dust Bowl: The Southern Plains in the 1930s* (New York: Oxford University Press, 1979).
12. Ibid., 4.
13. Ibid., 5.
14. Theodore Steinberg, *Nature Incorporated: Industrialization and the Waters of New England* (New York: Cambridge University Press, 1991).
15. Bonnie J. McCay and James Acheson, eds., *The Question of the Commons: The Culture and Ecology of Communal Resources* (Tucson: University of Arizona Press, 1987).
16. Garrett Hardin, "The Tragedy of the Commons," *Science* 162 (1968): 1243–48; Garrett Hardin and John Baden, *Managing the Commons* (San Francisco: W.H. Freeman, 1977).
17. Acheson, *Capturing the Commons.*
18. Ibid., 1.
19. Ibid., 2.
20. Richard W. Judd, *Aroostook : A Century of Logging in Northern Maine* (Orono: University of Maine Press, 1988).
21. Judd, *Common Lands, Common People*, 5.
22. Ibid., xii.

1. "WHITE-WASHED YANKEES": THE BEGINNINGS OF THE BAIT TRADE, 1790–1854

1. House of Assembly of Nova Scotia, "Letter from Philip Carten, Bryer's Island, 13 March 1837," *Journal of the Assembly*, appendix 75, 1837.
2. Ibid.
3. George W. Hilton, *The Truck System Including a History of the British Truck Acts, 1465–1960* (Westport, CT: W. Heffer, 1960), 6–8, 119–42; Ommer, *From Outpost to Outpost: A Structural Analysis of the Jersey-Gaspé Cod Fishery, 1767–1886* (Montreal: McGill-Queen's University Press, 1991); Daniel Vickers, *Farmers and Fishermen: Two Centuries of Work in Essex County, Massachusetts, 1630–1850* (Chapel Hill: University of North Carolina Press, 1994); Sean Cadigan, *Hope and Deception in Conception Bay: Merchant-Settler Relations in Newfoundland, 1785–1855* (Toronto: University of Toronto Press, 1995).
4. Shannon Ryan, "Fishery to Colony: A Newfoundland Watershed, 1793–1815," in *The Acadiensis*

Reader, vol. 1, 3rd ed., ed. P. A. Buckner, Gail G. Campbell, and David Frank (Fredericton, NB: Acadiensis Press, 1998), 177–78; Peter Pope, "The 16th-Century Fishing Voyage," in *How Deep is the Ocean? Historical Essays on Canada's Atlantic Fishery*, ed. James Candow and Carol Corbin (Sydney, NS: University College of Cape Breton Press, 1997), 15–30; Darlene Abreu-Ferreira, "Portugal's Cod Fishery in the 16th Century: Myths and Misconceptions," in *How Deep is the Ocean?*, 31–44; Jean-François Bière, "The French Fishery in North America in the 18th Century," in *How Deep is the Ocean?*, 45–64; Judith Tulloch, "The New England Fishery and Trade at Canso, 1720–1744," in *How Deep is the Ocean?*, 65–73.

5. Vickers, *Farmers and Fishermen*, 86–100.
6. For a thorough overview of the operations of the truck system, see Rosemary Ommer, ed., *Merchant Credit and Labour Strategies in Historical Perspective*, 10.
7. House of Assembly of Nova Scotia, "Letter from Gilbert Tucker, Montegan Clare, 11 March 1837," *Journal of the Assembly*, appendix 75, 1837.
8. For an overview of nineteenth-century British economic policies and the shift to a free-trade philosophy, see C. C. Eldridge, *England's Mission: The Imperialism Ideas in the Age of Gladstone and Disraeli, 1868–1880* (New York: MacMillan Press, 1973); D. K. Fieldhouse, *Economics and Empire, 1830–1914* (Ithaca, NY: Cornell University Press, 1973); John Gallagher and Ronald Robinson, "The Imperialism of Free Trade," *Economic History Review*, 2nd series, 6, no. 1 (1953): 1–15; J. A. S. Grenville, *Lord Salisbury and Foreign Policy: The Close of the Nineteenth Century* (London: University of London Press, 1964).
9. In the years just after the American Revolution, the West Indies consumed 60 percent of the American-caught fish, but by the 1840s the United States used 75 percent of its own catch from the North Atlantic. See Roger F. Duncan, *Coastal Maine: A Maritime History* (New York: W.W. Norton Press, 1992), 410–13; Lowell Armytage, *The Free Port System in the British West Indies: A Study in Commercial Policy, 1766–1833* (New York: Longmans, Green and Co., 1953); F. Lee Benns, *The American Struggle for the British West Indian Trade, 1815–1830* (Bloomington: University of Indiana Press, 1923).
10. House of Assembly of Nova Scotia, "A Message to Our Sovereign Lord George the Fourth," *Journal of the Assembly*, April 4, 1822, 206.
11. House of Assembly of Nova Scotia, "A Message to the King's Most Excellent Majesty," *Journal of the Assembly*, February 23, 1830, 597.
12. Ibid., 598.
13. House of Assembly of Nova Scotia, "A Message to the King's Most Excellent Majesty," 598.
14. The Assembly of Jamaica, as quoted in *St. Christopher Advertiser*, April 2, 1822.
15. House of Assembly of Nova Scotia, "A Message to Our Sovereign Lord George the Fourth," 206.
16. Stuart Bruchey, *Growth of the Modern American Economy* (New York: Dodd Mead, 1975), xi, 3, 9, 12, 35, 46, 63; Richard DuBoff, *Accumulation and Power: An Economic History of the United States* (Armonk, NY: ME Sharpe, 1989), 14–16.
17. House of the Assembly of Nova Scotia, "Report of Captain James Daly, January 14, 1853," *Journal of the Assembly*, appendix 4, 1853, 118.
18. Howard I. Chapelle, *The American Fishing Schooners, 1825–1935* (New York: W.W. Norton & Co., 1973).
19. Wayne O'Leary, *Maine Sea Fisheries: The Rise and Fall of a Native Industry* (Boston: Northeastern University, 1996), 41–71.
20. *Select Committee on the Origins of Fishing Bounties and Allowances, Majority Report*, 26th Cong., 1st sess., 1840, Senate Document 368, 4–25. See also O'Leary, *Maine Sea Fisheries*, 41–77.

21. For further information on regional economic interests, see Benjamin W. LaBaree, *America and the Sea: A Maritime History* (Mystic, CT: Mystic Seaport Publications, 1998), 387, 536, 541; and William W. Freehling, *The Road to Disunion: Secession at Bay, 1776–1854* (New York: Oxford University Press, 1990).
22. *Select Committee on the Origins of Fishing Bounties and Allowances, Majority Report*, 14.
23. Ibid., 25.
24. J. Ross Browne, *Report on the Bounty Allowances to Fishing Vessels*, report prepared for the Department of Treasury, 33rd Cong., 1st sess., November 25, 1853, Senate Document 3, 96–99.
25. The debate concerning the bounty system fell within the context of the larger debates regarding the role of the federal government and the state governments, as well as the growing sectional division between the North and the South. Any legislation proposed by New England representatives was sure to come under attack by southern politicians. For an example of this, see John Davis, *Select Committee on the Origins of Fishing Bounties and Allowances, Minority Report*, 26th Cong., 1st sess., 1840, Senate Document 368, *Minority Report*, 61; "Petition from John Williams and 87 Others from Kittery, February 14, 1839," as quoted in John Davis, *Select Committee on the Origins of Fishing Bounties and Allowances, Minority Report*, 64.
26. Browne, *Report on the Bounty Allowances to Fishing Vessels*, 88–89.
27. John Davis, *Select Committee on the Origins of Fishing Bounties and Allowances, Minority Report*, 61.
28. Ibid.
29. Ibid., 63.
30. Ibid.
31. "Petition from John Williams and 87 Others from Kittery, February 14, 1839," as quoted in John Davis, *Select Committee on the Origins of Fishing Bounties and Allowances, Minority Report*, 64.
32. John Anderson, *Documents in Relation to the Bounties Allowed to Vessels Employed in the Fisheries*, report prepared for the Committee on Finance in the U.S. Senate, 26th Cong., 2nd sess., 1841, Senate Document 148.
33. Senate, Speech of Hon. R. Williams of Maine in the Senate, 25th Cong., 3rd sess., *Congressional Globe* (January 29, 1839): 73–74.
34. Ibid.
35. Ibid.
36. Ibid.
37. O'Leary, *The Maine Sea Fisheries*, 75–77. See also House, Speech of Hon. Timothy Davis of Massachusetts in the House of Representatives, 34th Cong., 3rd sess., *Congressional Globe* (February 10, 1857): 218–22; House, Speech of Hon. N. Abbott of Maine in the House of Representatives, 35th Cong., 2nd sess., *Congressional Globe* (February 10, 1859): 102–4.
38. J. Murray Beck, *Nicholson-Fielding, 1710–1896*, vol. 1 of *Politics of Nova Scotia* (Tantallon, NS: Four East Publications, 1985), 23.
39. Ibid., 29.
40. Ibid., 43.
41. Ibid., 83.
42. House of Assembly of Nova Scotia, "Report of the Committee on the Subject of the Fisheries of the Province," *Journal of the Assembly*, February 24, 1824, 106.
43. Ibid., 107.
44. Ibid.

45. House of Assembly of Nova Scotia, "Proceedings," *Journal of the Assembly*, March 17, 1823, 261–62.
46. House of Assembly of Nova Scotia, "Proceedings," *Journal of the Assembly*, March 10, 1824, 406.
47. House of Assembly of Nova Scotia, "Report of the Committee on the Subject of the Fisheries of the Province, Presented by Mr. Harthshore," *Journal of the Assembly*, April 2, 1825, 512.
48. House of Assembly of Nova Scotia, "Proceedings," *Journal of the Assembly*, March 8, 1826, 594.
49. House of Assembly of Nova Scotia, "Report of the Committee on the Subject of the Fisheries of the Province, Presented by Mr. Fairbanks," *Journal of the Assembly*, March 16, 1827, 210–11.
50. House of Assembly of Nova Scotia, "Report of the Committee on the Subject of the Fisheries of the Province, Presented by Mr. Uniacke," *Journal of the Assembly*, February 24, 1829, 393. In the early nineteenth century, the advisory council for the executive of the colony largely controlled the government. The Crown's representative, not the elected assembly, appointed this council. The reform movement sought to replace this appointed council with one that represented the elected body. The idea was to make the executive responsible to the assembly.
51. House of Assembly of Nova Scotia, "Report of the Committee on the Subject of the Fisheries of the Province, Presented by Mr. Uniacke," 393.
52. House of Assembly of Nova Scotia, "Proceedings," *Journal of the Assembly*, April 6, 1833, 463; April 1, 1834, 647; April 11–12, 1834, 648.
53. House of Assembly of Nova Scotia, "Petition from John Taylor and Others," *Journal of the Assembly*, appendix 34, 1835; and "Petition No. 22 from Samuel Reynolds and Others," *Journal of the Assembly*, appendix 31, 1834.
54. House of Assembly of Nova Scotia, "Report of the Committee of the Fisheries," *Journal of the Assembly*, appendix 62, 103.
55. J. Murray Beck, *Nicholson-Fielding*, 19–67; Phillip A. Buckner, *The Transition to Responsible Government: British Policy in British North America, 1815–1850* (Westport, CT: Greenwood Press, 1985).
56. Beck, *Nicholson-Fielding*, 90, 93.
57. Ibid., 131.
58. "Responsible government" was a form of home rule that allowed the colony to control internal government activities such as taxation, education, health, transportation improvements, and elections. At the same time, the imperial authorities maintained control over international and transcolonial legislation, and agreements such as the Reciprocity Treaty of 1854.
59. Buckner, *The Transition to Responsible Government*, 291.
60. Ibid., 6.
61. Buckner, *The Transition to Responsible Government*, 18; Beck, *Nicholson-Fielding*, 101.
62. Buckner, *The Transition to Responsible Government*, 9.
63. Ibid., 65, 333.
64. Beck, *Nicholson-Fielding*, 105.
65. Ibid., 108–11.
66. Buckner, *The Transition to Responsible Government*, 306.
67. Buckner, *The Transition to Responsible Government*, 295–305.
68. House of Assembly of Nova Scotia, "Letter to Sir Peregrine Maitland," *Journal of the Assembly*, March 12, 1832, 306–7.
69. House of Assembly of Nova Scotia, "Petition of John Taylor and Others, 12 January 1835," *Journal of the Assembly*, appendix 34.

70. House of Assembly of Nova Scotia, "Petition of George Bell and Others," *Journal of the Assembly*, February 18, 1837, 59.
71. "Article I, Convention of 20 October 1818, between the United States of America and His Britannic Majesty," in Joseph Pope, *A Short Account of the Negotiations Which Resulted in the Convention of the 20th October, 1818, between Great Britain and the United States, Together with the Text of the Treaty* (Ottawa, 1907).
72. House of Assembly of Nova Scotia, "Proceedings," *Journal of the Assembly*, March 7, 1837, 29.
73. House of Assembly of Nova Scotia, "Petition of Fishermen of St. Margaret's Bay, Dover, and Prospect," *Journal of the Assembly*, February 21, 1837, 32. House of Assembly of Nova Scotia, "Petition of Clement Hubert and Others," *Journal of the Assembly*, February 27, 1832.
74. House of Assembly of Nova Scotia, "Letter from Gilbert Tucker, Montegan Clare, 11 March 1837," *Journal of the Assembly*, appendix 75, 1837.
75. House of Assembly of Nova Scotia, "Letter from Thomas Small," "Letter from James H. F. Randolph," "Letter from John Bass, Liverpool, 11 March 1837," "Letter from D & E Starr & Co., Halifax, 23 February 1837," "Letter from Joseph Allison & Co., New Glasgow, 11 March 1837," "Letter from William McLean, Pictou, 14 March 1837," "Letter from Thomas Tobin, Prospect, 15 March 1837," "Letter from Locke & Churchill, Rugged Island, 13 March 1837," *Journal of the Assembly*, appendix 75, 1837.
76. House of Assembly of Nova Scotia, "Letter from Thomas Small," *Journal of the Assembly*, appendix 75, 1837.
77. House of Assembly of Nova Scotia, "Report of Captain Alexander Miline, of the HMS *Crocodile* to Vice Admiral Sir Thomas Harvey, KCB, Commander in Chief, & c.," *Journal of the Assembly*, appendix 62, 1854, 173.
78. House of Assembly of Nova Scotia, "Letter from Joseph Allison & Co., New Glasgow, 11 March 1837," *Journal of the Assembly*, appendix 75, 1837.
79. House of Assembly of Nova Scotia, "Letter from D & E Starr & Co., Halifax, 23 February 1837," *Journal of the Assembly*, appendix 75, 1837.
80. House of Assembly of Nova Scotia, "Report of Commander F. Egerton, of the HMS *Basilisk*," *Journal of the Assembly*, appendix 2, 1854, 20.
81. House of Assembly of Nova Scotia, "Letter from Elisha Payson, Bryer's Island, 13 March 1837," *Journal of the Assembly*, appendix 75, 1837.
82. House of Assembly of Nova Scotia, "Letter from Paul Crowell to James B. Uniacke, 10 February 1852," *Journal of the Assembly*, appendix 25, 1852, 171–72.
83. House of Assembly of Nova Scotia, "Report of M.R. Perchell, HMS *Alice Rogers*, 23 October 1853," *Journal of the Assembly*, appendix 2, 1854, 27. See also "Report of A.F. DeHorsey, HMSS *Devastation*, 23 October 1853," *Journal of the Assembly*, appendix 2, 1854, 35. DeHorsey reported that "three-fourths, or at least one-half of the crews of vessels under American colors are our own countrymen."
84. House of Assembly of Nova Scotia, "Report of the Committee of the Fisheries, 10 April 1837, James Uniacke, Chair," *Journal of the Assembly*, appendix 75, 1837, 4.
85. House of Assembly of Nova Scotia, "Report of M.R. Perchell, HMS *Alice Rogers*, 23 October 1853," *Journal of the Assembly*, appendix 2, 1854, 27.
86. House of Assembly, Nova Scotia, "Report of Lt. W.W. Bridges, Armed Tender *Bonito*, October 20, 1853," *Journal of the Assembly*, appendix 2, 1854, 29.
87. House of Assembly of Nova Scotia, "Letter from Thomas Tobin, March 15, 1837," *Journal of the Assembly*, appendix 75, 1837.

88. These reforms were part of the larger political reform movements in Nova Scotia during the middle decades of the nineteenth century. See David Alexander, "New Notions of Happiness: Nationalism, Regionalism and Atlantic Canada," *Journal of Canadian Studies* 15 (Summer 1980): 29–42; Phillip A. Buckner and David Frank, eds., *Atlantic Canada before Confederation* (Fredericton, NB: Acadiensis Press, 1990); Buckner, *The Transition to Responsible Government*; Phillip A. Buckner and John G. Reid, eds., *The Atlantic Region to Confederation: A History* (Toronto: University of Toronto, 1994).
89. House of Assembly of Nova Scotia, "Letter from William Crichton, 20 March 1837," *Journal of the Assembly*, appendix 75, 1837.
90. House of Assembly, Nova Scotia, "Letter from A.D. Gordon, Pictou, 14 March 1837," *Journal of the Assembly*, appendix 75, 1837 (italics in original text).
91. House of Assembly of Nova Scotia, "Report of Colin York Campbell, HMSS *Devastation*, 10 November 1852," *Journal of the Assembly*, appendix 4, 1853, 115.
92. Lorenzo Sabine, *Report on the Principal Fisheries of the American Seas*, prepared for the Department of Treasury, 32nd Cong., 2nd sess., 1853., Sen. Exec. Doc. 22, vol. 6, serial 662, 365.
93. House of Assembly of Nova Scotia, "Letter from Thomas Small," "Letter from Joseph Allison & Co., New Glasgow, March 11, 1837," "Letter from Gilbert Tucker, Montegan Clare, 11 March 1837," "Letter from Locke & Churchill, Rugged Island, 13 March 1837," "Letter from C. McAlpine," "Letter from William Crichton, 20 March 1837," *Journal of the Assembly*, appendix 1837.
94. House of Assembly of Nova Scotia, "Letter from Lieutenant-General Sir Colin Campbell, 5 November 1838," *Journal of the Assembly*, appendix 9, 1838.
95. "Letter from Lord Glenelg, Downing Street, dated 5 November 1836, re: fisheries, to Sir Colin Campbell, Lieut. General of Nova Scotia," Nova Scotia Archives and Records Management, MG 100, vol. 146, no. 12; "Seizure of American Fishing Vessel," *The Nova Scotia*, August 9, 1852, 250, cols. 1–2, vol. 13, no. 32.
96. House of Assembly of Nova Scotia, "Report of the Committee on the Fisheries," *Journal of the Assembly*, appendix 62, 1841, 161. See also "Report of the Committee on the Fisheries," *Journal of the Assembly*, appendix 74, 1842, 184.
97. House of Assembly of Nova Scotia, "Report of the Committee on the Fisheries," *Journal of the Assembly*, appendix 75, 1847, 283. See also "Report of the Committee on the Fisheries," *Journal of the Assembly*, appendix 89, 1848, 70.
98. House of Assembly of Nova Scotia, "Report of the Committee on the Fisheries," *Journal of the Assembly*, appendix 73, 1851, 247.
99. House of the Assembly of Nova Scotia, *Journal of the Assembly*, appendix 4, 1853, 109–10.
100. Innis, *The Cod Fisheries*, 334–42.
101. For an overview of nineteenth-century British economic policies and the shift to a free-trade philosophy, see Eldridge, *England's Mission*; Fieldhouse, *Economics and Empire*; Grenville, *Lord Salisbury and Foreign Policy*; John Gallagher and Ronald Robinson, "The Imperialism of Free Trade," *Economic History Review*, 2nd series, 6, no. 1 (1953): 1–15.
102. House of Assembly of Nova Scotia, "Letter of James Laybold, Halifax, November 30, 1852," *Journal of the Assembly*, appendix 4, 1853, 122. See also House of Assembly of Nova Scotia, "Reports of F. Egerton, Commander of HMS *Basilisk*, 11 November 1853," "Report of A.F. DeHorsey HMSS *Devastation*, 28 October 1853," "Report of Lt. J. Jenkins HMS *Cumberland*, 8 November 1853," "Report of Lt. M.R. Perchell HM Armed Tender *Alice Rogers*, 23 October 1853," "Report of Lt. C.G. Linday HM Armed Tender *Bonito*, 8 August 1853," and "Report

of Lt. W.W. Bridges HM Armed Tender *Bonito*, 23 October 1853," *Journal of the Assembly*, appendix 2, 1854, 20–29.

103. For histories of the 1854 Reciprocity Treaty, see R. E. Ankli, "Canadian-American Reciprocity: A Comment," *Journal of Economic History* 28 (June 1970): 274–81; R. E. Ankli, "The Reciprocity Treaty of 1854," *Canadian Journal of Economic History* 4 (1971): 1–20; Charles Campbell, *From Revolution to Rapprochement: The United States and Great Britain, 1783–1900* (New York: Wiley, 1974); Julian Gwyn, "Tariffs, Trade, and Reciprocity: Nova Scotia, 1830–1866," *Acadiensis* 25, no. 2 (Spring 1996): 62–81; Julian Gwyn, "Golden Age or Bronze Age? Wealth and Poverty in Nova Scotia: The 1850s and 1860s," in *Canadian Papers in Rural History*, vol. 3, ed. Donald H. Akensen (Gananoque, ON: Langdale Press, 1992); Irene Hecht, "Israel D. Andrews and the Reciprocity Treaty of 1854: A Reappraisal," *Canadian Historical Review* 44 (December 1963): 313–26; R. H. McDonald, "Nova Scotia and the Reciprocity Negotiations, 1845–1854: A Re-Interpretation," *Nova Scotia Historical Quarterly* (1977): 205–34; Donald C. Masters, *Reciprocity, 1846–1911* (Ottawa: Canadian Historical Association, Booklet 12, 1961); L. Officer and L. Smith, "The Canadian American Reciprocity Treaty of 1855 to 1868," *Journal of Economic History* 28 (1968): 598–623; Gregory A. Raymond, "Canada between the Superpowers: Reciprocity and Conformity in Foreign Policy," *American Review of Canadian Studies* 17 (Summer 1987): 221–36; S. A. Saunders, "The Maritime Provinces and the Reciprocity Treaty," *Dalhousie Review* 14 (October 1934): 22–43; S. A. Saunders, "The Reciprocity Treaty of 1854: A Regional Study," *Canadian Journal of Economic and Political Science*, no. 2 (February 1936): 41–53.
104. Wayne O'Leary, *Maine Sea Fisheries*, 248–51.

2. "INTRUSION OF STRANGERS": SEEKING LOCAL CONTROL IN AN EMERGING NATIONAL CONTEXT, 1854–1885

1. George Brown Goode, *The Fisheries and Fishing Industry of the United States*, vol. 5, *History and Method of the Fisheries* (Washington, DC: Government Printing Office, 1887), 446.
2. Howard Jones, *Union in Peril: The Crisis over British Intervention in the Civil War* (Chapel Hill: University of North Carolina, 1992).
3. John Herd Thompson and Stephen J. Randall, *Canada and the United States: Ambivalent Allies*, 2nd ed. (Athens: University of Georgia Press, 1997).
4. Nell Irvin Painter, *Standing at Armageddon: The United States, 1877–1919* (New York: W.W. Norton & Company, 1987), 73–83, 115, 253–56, 272; Lewis L. Gould, "Party Conflict: Republicans versus Democrats, 1877–1901," in *The Gilded Age: Perspectives on the Origins of Modern America*, 2nd ed., ed. Charles Calhoun (Lanham, MD: Rowman & Littlefield Publishers, 2007), 266, 276.
5. "The Reciprocity Treaty—The Fishery Question," *Baltimore Sun*, March 8, 1866.
6. Some examples include "The Fishery Question Its Previous History–Breakers Ahead," *Memphis Daily Avalanche*, March 7, 1866, 4; "The Fishery Question," *(Galveston, TX) Flake's Bulletin*, March 20, 1866, 4; "The Fishery Question," *San Francisco Evening Bulletin*, March 31, 1866, 2.
7. E. R. Forbes and D. A. Muise, eds., *The Atlantic Provinces in Confederation* (Toronto: University of Toronto Press, 1993); Jonathan Swainger, *The Canadian Department of Justice and the Completion of Confederation, 1867–78* (Vancouver: University of British Columbia Press, 2000); Frederick Vaughan, *The Canadian Federalist Experiment: From Defiant Monarch to Reluctant*

Republic (Montreal: McGill-Queen's Press, 2003); P. B. Waite, *The Life and Times of Confederation, 1864–1867: Politics, Newspapers and the Union of British North America*, 3rd ed. (Toronto: University of Toronto Press, 1962).

8. "The Fisheries," *Halifax Evening Reporter*, February 2, 1866, 1, col. 4.
9. "The Fishery Question Its Previous History–Breakers Ahead," *Memphis Daily Avalanche*, March 7, 1866, 4.
10. "The Reciprocity Treaty—The Fishery Question," *Baltimore Sun*, March 8, 1866.
11. "The Fishery Question," *San Francisco Evening Bulletin*, March 31, 1866, 2; "Telegrams: The Newfoundland Fishery Question," *Salt Lake Daily Telegraph*, April 1, 1866, 2; "The Fishery Question," *(Washington, DC) Daily Constitutional Union*, April 14, 1866, 2.
12. "The Blunder about Canada: The Fenians and the Fisheries," *New Orleans Daily Picayune*, April 27, 1866, 2. See also "Reciprocity and the Fisheries," *Philadelphia Inquirer*, June 1, 1866, 4.
13. "The Canadian Parliament," *Albany Journal*, August 11, 1866, 3; "The Fishery Dispute," *Philadelphia Inquirer*, June 6, 1866, 4.
14. "Annual Report of the Department of Marine and Fisheries, for the Fiscal Year Ended 30th, June, 1869," *Sessional Papers*, no. 11 (1870): 49–52.
15. "Mr. Seward's Diplomacy," *Troy Weekly Times*, May 16, 1868, 2.
16. "Appendix No. 12—Report of W.H. Venning, Esq., Inspector of Fisheries, New Brunswick and Nova Scotia, for 1869," in "Annual Report of the Department of Marine and Fisheries, for the Fiscal Year Ended 30th, June, 1869," *Sessional Papers*, no. 11 (1870): 79–80, 101.
17. Canada, *Parliamentary Debates, 1870*, March 3, 1870, as quoted in Ronald D. Tallman, "Warships and Mackerel: The North Atlantic Fisheries in Canadian-American Relations, 1867–1877" (Ph.D. diss., University of Maine, 1971), 328–29.
18. Tallman, "Warships and Mackerel," 146–9. See also Andrew Hill Clark, *Three Centuries and the Island: A Historical Geography of Settlement and Agriculture in Prince Edward Island, Canada* (Toronto: University of Toronto Press, 1959).
19. Letter from Cardwell, Secretary of State for the Colonies to the Lieutenant Governor of Nova Scotia, May 26, 1866, as quoted in North Atlantic Coast Fisheries, *Part I of the Appendix to the Case of the United States Including Treaties, Statutes, and Correspondences*, vol. 2 of *Proceedings in the North Atlantic Coast Fisheries Arbitration before the Permanent Court of Arbitration at The Hague* (Washington, DC: Government Printing Office, 1912), 152–53.
20. "Lt. Cochrane, November 18, 1870," as quoted in Bernard Bayard, *Selected Cases of Maltreatment of American Vessels*, appendix B of *Record of the Proceedings of the Fisheries Conference at Washington, Nov. 1887*, Sir Charles Tupper Papers, Political Correspondences, under Memoranda and Report, reel 10. See also other printed materials, reels 10 and 11, Microfilm 11,110, Nova Scotia Archives and Records Management.
21. "Report of Peter Mitchell, Annual Report of the Department of Marine and Fisheries for the Fiscal Year Ended 30th June, 1869," *Sessional Papers*, no. 11 (1870).
22. "Letter, from Foreign Office to Secretary of the Admiralty, 30 April 1870, included in North Atlantic Coast Fisheries," *Part I of the Appendix to the Case of the United States*, vol. 2, 143.
23. "Letter from the Lords of Admiralty to Vice-Admiral Wellesley, 5 May 1870, included in North Atlantic Coast Fisheries," *The Case of the United States*, vol. 1 of North Atlantic Coast Fisheries Arbitration, *Proceedings in the North Atlantic Coast Fisheries Arbitration before the Permanent Court of Arbitration at The Hague* (Washington, DC: Government Printing Office, 1912), 143.
24. "Memorandum of Peter Mitchell, Ottawa, 31 May 1870, in Report of the Minister for the

Fourth Annual Report of the Department of Marine and Fisheries for the Year 1871," *Sessional Papers*, no. 5 (1872), 60.

25. "Letter from Granville to Young, Colonial Office, 30 June 1870," in Bernard Bayard, *Selected Cases of Maltreatment of American Vessels.*
26. "Letter from Kimberly to Lisgar, 17 March 1871," in Bernard Bayard, *Selected Cases of Maltreatment of American Vessels.*
27. Ibid.
28. "Fourth Annual Report of the Department of Marine and Fisheries, Being for the Fiscal Year Ended 30th, June, 1871," *Sessional Papers*, no. 5 (1872): 60.
29. *Montreal Herald*, as quoted in the *Canadian News*, April 14, 1870.
30. "R. Hodgson to Right Hon. Earl Granville, K.G., Secretary of State for the Colonies, June 1, 1870," in *Correspondences between the Lieutenant Governor and Her Majesty's Principal Secretary of State for the Colonies, on the Subject of the Fisheries*, no. 3 of *Correspondences on the Subject of Admitting United States Fishing Vessels to Entry in the Ports of Prince Edward Island*, appendix K of the Journal of the House of Assembly, Prince Edward Island, 1871.
31. "Council Chamber to His Honor Sir Robert Hodgson, Knight, Administrator of the Government of Prince Edward Island, in Council, September 2, 1870," in no. 1 of *Correspondences on the Subject of Admitting United States Fishing Vessels to Entry in the Ports of Prince Edward Island*, appendix K of the Journal of the House of Assembly, Prince Edward Island, 1871.
32. "Mr. Collector Clark to Sir R. Hodgson, August 25, 1870," in *Correspondences between Vice-Admiral Wellesley, Captain Hardinge, Commander Poland, and Sir Robert Hodgson, Relative to the Violation of the Fishery Treaty and Laws*," no. 2 of *Correspondences on the Subject of Admitting United States Fishing Vessels to Entry in the Ports of Prince Edward Island*, appendix K of the Journal of the House of Assembly, Prince Edward Island, 1871.
33. "Captain Hardinge to Sir R. Hodgson, Relative to Schooner *Clara B. Chapman*, August 20, 1870," ibid.
34. "Captain Hardinge to Sir R. Hodgson, August 20, 1870."
35. "Mr. Collector Clark to Sir R. Hodgson, August 25, 1870."
36. "Mr. Collector Clark to Sir R. Hodgson, August 25, 1870," and "Captain Hardinge to Sir R. Hodgson, August 20, 1870."
37. "Mr. Collector Clark to Sir R. Hodgson, August 31, 1870."
38. "Captain Hardinge to Sir R. Hodgen, reporting irregularity in Charlottetown Harbor on the part of the Steamer 'Georgia,' August 27, 1870."
39. "Extracts from the Minutes of the Executive Council of Prince Edward Island, September 2, 1870."
40. "Extracts from the Minutes of the Executive Council of Prince Edward Island, September 2, 1870"; "Assistant Colonial Secretary to Captain Hardinge, August 31, 1870."
41. "The Fishery Business Again," *Charlottetown Patriot*, August 27, 1870; November 10, 1870.
42. Ibid.
43. For an overview of Ben Butler's attitude and politics, see Tallman, "Warships and Mackerel," 251–54.
44. *Boston Herald*, June 20, 1870.
45. For example, see "General News: British Circular Regarding Fisheries—The Three Mile Line," *San Francisco Evening Bulletin*, July 18, 1870, 4.
46. "The Fishery Dispute," *(Philadelphia) North American*, November 3, 1870, 1.
47. As quoted in "Canada: Further Comments on Grant's Message," *Cincinnati Daily Gazette*,

December 9, 1870, 3.

48. "The President's Message," *Galveston Daily News*, December 9, 1870, 1. See also "The British Questions," *Cincinnati Daily Gazette*, December 29, 1870, 2, which states: "This onus is on us, and until we have made a rational effort we have no complaint against Great Britain; and when, in this situation, we use offensive language, and assume that there is an irreconcilable dispute, we merely show the Butlerian policy of making a foreign quarrel to prevent a free election at home."
49. "From Washington, Scenes at the Opening of Congress—The Tug over Revenue Reform," *Springfield (MA) Republican*, December 9, 1870, 4.
50. "The British Questions," *Cincinnati Daily Gazette*, December 29, 1870, 2.
51. *Cape Ann Advertiser*, August 26, 1870.
52. For an overview of the exchanges between the American and Canadian newspapers, see "Canada," *Cincinnati Daily Gazette*, March 3, 1870, 2. This paper referred to the *New York Tribune* as "the head lunatic of the prohibitory policy."
53. *(Saint John) Morning Freeman*, December 15, 1870.
54. *New York Times*, March 17, 1870; and *(Washington, DC) Daily Republican*, October 21, 1870.
55. Court of Vice-Admiralty, *Records of the Court of Vice-Admiralty at Halifax, Nova Scotia*, Nova Scotia Archives and Records Management, Record Group 40, vol. 25, files 7, 25, and 26.
56. "General Report for 1870, EG Fanshawe, Vice-Admiral, *Royal Alfred*, Halifax, 22 Nov 1870," in "Correspondences between the Government of the Dominion and the Imperial Government on the Subject of the Fisheries, with Other Documents Relating to the Same, Laid before the Honorable House of Commons 34 Victoria," *Sessional Papers*, no. 12a (1871); "The Fishery Question," *Philadelphia Inquirer*, December 31, 1870, 1.
57. Tallman, "Warships and Mackerel," 259–65.
58. Tallman, "Warships and Mackerel," 299–314; "The Alabama Claims: A Joint Commission Appointed to Settle Them. It Will Sit in Washington," *Cincinnati Daily Gazette*, February 10, 1871, 3; "The Peace Commission," *(Macon) Georgia Weekly Telegraph*, February 21, 1871, 8. See also "How the Codfish Quarrel Re-Opened—Diplomacy on the Alabama Question," *(Macon) Georgia Weekly Telegraph*, February 21, 1871, 2, which fully outlines the shift in Grant's policy towards Canada as it related to other international issues and domestic political troubles. This paper also suggests that it was the British who appeared more eager to finally deal with the situation, along with many other conflicts in Anglo-American relations.
59. Thompson and Randall, *Canada and the United States*, 39–40.
60. C. P. Stacey, *Canada and the Age of Conflict*, vol. 1, *1867–1921* (Toronto: University of Toronto Press, 1984), 21–22.
61. *Correspondence with the Government of Canada in Connection with the Appointment of the Joint Commission and the Treaty of Washington: Presented to Both Houses of Parliament by Command of Her Majesty, April, 1872* (London: W. Clowes & Son, for H.M.S.O., 1872).
62. Tallman, "Warships and Mackerel," 329–38.
63. Ibid., 341–42.
64. *Halifax Chronicle*, as reported in the *(Portland, ME) Eastern Argus*, May 23, 1871.
65. Halifax Commission, *Records of the Proceedings of the Halifax Commission, 1877 under the Treaty of Washington of May 8, 1871* (Washington, DC: Government Printing Office, 1878).
66. "Appendix A: Case of Her Majesty's Government, Part I: Canada," *Records of the Proceedings of the Halifax Commission, 1877*.
67. Ibid.

68. Spencer Baird, "Report of the Commissioner for 1877," in *Part V, Report of the Commissioner for 1877* (Washington, DC: Government Printing Office, 1877), 12–13.
69. "Appendix L—United States Evidence," and "Appendix M—United States Affidavits," *Records of the Proceedings of the Halifax Commission, 1877*.
70. "Appendix O—United States Statistics," *Records of the Proceedings of the Halifax Commission, 1877*.
71. "Volume I: Protocols and Decision of the Proceedings," *Records of the Proceedings of the Halifax Commission, 1877*.
72. "Affidavit of Peter McAulay," appendix B of no. 347 in U.S. Department of State, *Papers of the Foreign Relations of the United States* (Washington: Government Printing Office, 1879), 560.
73. "Affidavit of David Malanson," appendix A of no. 375 in U.S. Department of State, *Papers of the Foreign Relations of the United States*, 584.
74. "Affidavit of Joseph Bowie," no. 301, enclosure 2 in U.S. Department of State, *Papers of the Foreign Relations of the United States*, 499.
75. "Affidavit of John Dago," no. 301, enclosure 1 in U.S. Department of State, *Papers of the Foreign Relations of the United States*, 498.
76. "Affidavit of Mark Bolt," no. 375 in U.S. Department of State, *Papers of the Foreign Relations of the United States*, 576.
77. "Affidavit of Richard Henriken," no. 375 in U.S. Department of State, *Papers of the Foreign Relations of the United States*, 577. See also Sean T. Cadigan, "Failed Proposals for Fisheries Management and Conservation in Newfoundland, 1855–1880," in *Fishing Places, Fishing People: Issues in Small Scale Fisheries*, ed. Dianne Newell and Rosemary Ommer (Toronto: University of Toronto Press, 1998); Cadigan, "The Moral Economy of the Commons: Ecology and the Equity in Newfoundland Cod Fisheries, 1815–1855," *Labour/Le Travail* 43 (Spring 1999): 9–42.
78. "American Fishermen and Canadian Fisheries Regulations," *Novascotian*, July 23, 1887, 8.
79. "Report of William Wakeham, the Fishery Officer in Charge of the Fisheries Protection Service in the Gulf and Lower St. Lawrence, Appendix No. 1 to Supplement No. 2 to the Twelfth Annual Report of the Department of Marine and Fisheries for the Year 1879. Report on the Fisheries of Canada for the Year 1879," *Sessional Papers*, no. 9 (1880), 81.
80. "Report of William Wakeham, the Fishery Officer in Charge of the Fisheries Protection Service in the Gulf and Lower St. Lawrence, during the Season of 1880, Appendix No. 3 of Supplement No. 2 of the Thirteenth Annual Report of the Department of Marine and Fisheries, Year Ending 30th June, 1880," *Sessional Papers*, no. 11 (1881), 77.
81. "Report of William Wakeham on the Gulf of St. Lawrence Fisheries for the Year 1895, Including Synopsis of the Local Overseers' Reports, Appendix No. 6 of the Twenty-Ninth Annual Report of the Department of Marine and Fisheries, 1896," *Sessional Papers*, no. 11a (1897): 153–54.
82. "Report of the Marine Police Vessel *Peter Mitchell* Commanded by D.M. Browne, Navigating Lieutenant, RN, Appendix X of the Fifth Annual Report of the Department of Marine and Fisheries, Being for the Fiscal Year Ended 30th, June, 1872," *Sessional Papers*, no. 8 (1873): 202.
83. Phillip A. Buckner, "The 1870s: Political Integration," in *The Atlantic Provinces in Confederation*, ed. E. R. Forbes and D. A. Muise, 80.
84. Vaughan, *The Canadian Federalist Experiment*, 63–67.
85. Swainger, *The Canadian Department of Justice and the Completion of Confederation*, 5.
86. Stacey, *Canada in the Age of Conflict*, 32.

87. Swainger, *The Canadian Department of Justice and the Completion of Confederation*, 14–18.
88. For a review of Macdonald's political philosophy and the influence of Thomas Hobbes's ideas of a strong central power in government upon Canadian politics of the 1860s and 1870s, see Vaughan, *The Canadian Federalist Experiment*, 62, 88–89, 91–113. See also Allan Greer and Ian Ranforth, eds., *Colonial Leviathan: State Formation in Mid-Nineteenth-Century Canada* (Toronto: University of Toronto Press, 1992).
89. Judith Fingard, "The 1880s: Paradoxes of Progress," in *The Atlantic Provinces in Confederation*, ed. E. R. Forbes and D. A. Muise, 90–101.

3. "A FISHERMAN OUGHT TO BE A FREE TRADER ANYWAY": THE BAIT TRADE IN DIPLOMATIC CONTROVERSY, 1886–1888

Quote from "Great Britain and the Fisheries," *Novascotian*, March 13, 1886, 1, as part of the newspaper's attempt to convince the fishing communities of Nova Scotia to support the Liberal party in the next election and vote against the "protectionists" within John Macdonald's Conservative party.

1. "Arrival of the Str. Lansdowne: In Hot Pursuit of American Fishermen," *Digby Weekly Courier*, May 7, 1886; Petition of Wallace Graham, Solicitor of the Attorney General of Canada, in Case file 427, "Case File for the Queen v. *David J. Adams*," Vice-Admiralty Court, Halifax, Nova Scotia, Nova Scotia Archives and Records Management, RG 40, no. 42, folders 3–4.
2. Examination of Edwin C. Dodge, master mechanic, Digby, foreman of Government pier at Digby, Examination of Robert Spurr, of Clements, and Examination of Owen Riley, Digby, in Case file 427, "Case File for the Queen v. *David J. Adams*."
3. "The First Bona Fide Seizure: The Fisherman David J. Adams Captured at Digby," *Halifax Morning Herald*, May 8, 1886, 2.
4. Oath of Peter A. Scott, Halifax, Fishery Officer of the Dominion of Canada, in Case file 427, "Case File for the Queen v. *David J. Adams*"; "The Captured Fisherman," *Halifax Morning Herald*, May 10, 1886, 2.
5. "The Captured Fisherman," *Halifax Morning Herald*, May 10, 1886.
6. Halifax Commission, *Records of the Proceedings of the Halifax Commission, 1877 under the Treaty of Washington of May 8, 1871* (Washington, DC: Government Printing Office, 1878).
7. "Appendix L—United States Evidence," and "Appendix M—United States Affidavits," *Records of the Proceedings of the Halifax Commission, 1877*; "Appendix O—United States Statistics," *Records of the Proceedings of the Halifax Commission, 1877*.
8. Charles Levi Woodbury, *The Relations of the Fisheries to the Discovery and Settlement of North America; Delivered before the New Hampshire Historical Society, at Concord, June, 1880, and the Massachusetts Fish and Game Protection Society, at Boston, 1880* (Boston: Alfred Mudge & Son, Printers, 1880); Charles Isham, *The Fishery Question: Its Origin, History, & Present Situation: With a Map of the Anglo-American Fishing Grounds & a Short Bibliography* (New York: Putnam's Sons, 1887); "The Fisheries Question at Washington," *Digby Weekly Courier*, December 12, 1885.
9. Spencer F. Baird, United States Commission of Fish and Fisheries, *Report on the Condition of the Sea Fisheries of the South Coast of New England in 1871 and 1872* (Washington, DC: Government Printing Office, 1873); United States Commission of Fish and Fisheries, *Report of the Commissioner for 1877* (Washington, DC: Government Printing Office, 1877).

10. A letter from Boston fish merchant William Jones was published in "Correspondence: Reciprocity," *Novascotian*, July 18, 1885, 1.
11. "Reciprocity," *Novascotian*, June 20, 1885, 1; "The Agreement," *Novascotian*, July 4, 1885, 1.
12. George Brown Goode, *The Fisheries and Fishing Industry of the United States*, vol. 5, *History and Method of the Fisheries* (Washington, DC: Government Printing Office, 1887), 247–64. See also an interview with Spencer Baird, director of the U.S. Commission of Fish and Fisheries in the *New York Tribune*, January 21, 1886, in which he states that the privilege of fishing in Canadian waters has decreased in value since the American fleet moved farther offshore; for the Canadian response to this claim, see "The Fishery Question," *Novascotian*, January 30, 1886, 5.
13. "From Washington," *Boston Herald*, January 21, 1886; "Some Points about the Fisheries," *Novascotian*, January 30, 1886, 2.
14. "Article I, Convention of 20 October 1818, between the United States of America and His Britannic Majesty," in Joseph Pope, *A Short Account of the Negotiations Which Resulted in the Convention of the 20th October, 1818, between Great Britain and the United States, Together with the Text of the Treaty* (Ottawa, 1907).
15. "Protecting the Fisheries: A Talk with Capt. Scott, R.N." *Digby Weekly Courier*, March 5, 1886.
16. "Protecting Our Fisheries," *St. John Globe*, March 19, 1886. Emphasis in original.
17. *Yarmouth Times*, March 19, 1886.
18. *Digby Weekly Courier*, March 19, 1886.
19. "Local Parliament," *Digby Weekly Courier*, March 5, 1886.
20. "National Rights," *Digby Weekly Courier*, April 9, 1886. See also "Protecting the Fisheries," *St. John Globe*, March 19, 1886; *Charlottetown Herald*, July 1, 1886.
21. *Novascotian*, February 14, 1885, 1.
22. "The Fishery Treaty," *Novascotian*, January 10, 1885, 6; "Reciprocity," *Novascotian*, June 20, 1885, 1.
23. "That Agreement," *Novascotian*, July 4, 1885, 1.
24. *Novascotian*, September 19, 1885, 2; "How Much They Want for Reciprocity," *Novascotian*, October 24, 1885, 2; "Our Fisheries," *Novascotian*, May 1, 1886, 5.
25. "The Fisheries," *Novascotian*, January 23, 1886, 1. The *Novascotian* referred to Macdonald as "Old To-morrow" for his failure to take on the reciprocity issue as soon as the United States announced the repeal of the fishery clauses of the 1871 treaty. See "What Gloucester Got," *Novascotian*, February 6, 1886, 1; *Lunenburg County Times*, April 14, 1886.
26. "The Fisheries," *Novascotian*, January 23, 1886, 1; "The Fishery Treaty: Mr. Blake Refers to Reciprocity in the House of Commons," *Novascotian*, February 7, 1885, 6; "Some Points about the Fisheries," *Novascotian*, January 30, 1886, 2.
27. "The Fisheries: Significant Action of the Port Maitland Folks," *Novascotian*, April 3, 1886, 8.
28. "Key to the Whole Position: If We Yield on the Bait, We Yield Everything—Views of His Honor Judge Savary," *Halifax Herald*, June 4, 1886.
29. As quoted in "Some Points about the Fisheries," *Novascotian*, January 30, 1886, 2.
30. Ibid. See also *(Montreal) Journal of Commerce*, February 20, 1886.
31. "Another Boston View of the Fishery Question," *Novascotian*, March 6, 1886, 2.
32. "Great Britain and the Fisheries," *Novascotian*, March 13, 1886, 1.
33. *New York Herald*, April 4, 1886.
34. "Disgraceful Conduct of American Fishermen: Wanton Destruction of Property," *Digby Weekly Courier*, April 16, 1886; also covered in "The Fisheries," *Novascotian*, April 17, 1886, 5.
35. "Arrival of the Str. Lansdowne: In Hot Pursuit of American Fishermen," *Digby Weekly Courier*,

May 7, 1886.

36. "Cruise of the Lansdowne: American Fishermen Ordered off the Coast by Captain Scott," *Digby Weekly Courier*, April 2, 1886.
37. "Permit to Touch and Trade, US Port of Portland, April 24, 1886," in the Records of the Court of Vice-Admiralty, The Queen v. the Ship *Ella M. Doughty* (Vice-Adm. Ct. 1886), file 473, Nova Scotia Archives and Records Management, RG 40, vol. 43, no. 1.
38. "Testimony of Captain Warren A. Doughty, June 2, 1887," in the Records of the Court of Vice-Admiralty, The Queen v. the Ship *Ella M. Doughty.*
39. "Testimony of Donald McRitchie, Englishtown," "Testimony of Torqueil McLean, Englishtown," "Testimony of Malcolm McDonald, Englishtown," "Testimony of Donald McInnes, Englishtown," "Testimony of Donald J. Morrision, Englishtown," "Testimony of Daniel McAskell," and "Testimony of Angus McLeod, St. Ann's," in the Records of the Court of Vice-Admiralty, The Queen v. the Ship *Ella M. Doughty.*
40. "Testimony of Angus McLeod, St. Ann's," in the Records of the Court of Vice-Admiralty, The Queen v. the Ship *Ella M. Doughty.*
41. Thomas Bayard, *Selected Cases of Maltreatment of American Fishing Vessels* (Washington, DC: Government Printing Office, 1887), 2–3.
42. "Letter from Sec. of State T. F. Bayard to Hon. George F. Edmunds, US Senate, Committee on Foreign Relations, January 26, 1887," in *Correspondence Related to the North American Fisheries: 1884–86, Presented to Both Houses by Command of Her Majesty, February 1887* (Printed by Her Majesty's Stationery Office, London, by Harrison & Sons, 1887).
43. Letter from Sec. of State T. F. Bayard to Hon. George F. Edmunds, US Senate, Committee on Foreign Relations, January 26, 1887; "The Highland Light To Be Sold To-Day—Georgetown People Needlessly Excited," *Novascotian*, December 18, 1886, 3; "Highland Light Sold: Bought for the Government for $5,800, for Use as a Cruiser," *Novascotian*, December 18, 1886, 6.
44. Letter from Sec. of State T. F. Bayard to Hon. George F. Edmunds, US Senate, Committee on Foreign Relations, January 26, 1887.
45. "American Fishermen Seized at Shelburne for Violating Custom Laws," *Digby Weekly Courier*, July 9, 1886.
46. Petition of Wallace Graham, Solicitor of the Attorney General of Canada, in Case file 427, "Case File for the Queen v. *David J. Adams.*"
47. For a review of the case and counter-cases presented in the Vice-Admiralty Court, see "The Adams Seizure: Arguments of the Canadians and United States in the Admiralty Court," *Novascotian*, June 11, 1887.
48. Defense of Jessie Lewis (signed August 7, 1886, N. H. Meagher), in Case file 427, "Case File for the Queen v. *David J. Adams.*"
49. Examination of James B. Hill, Chief Officer of the *Lansdowne*, Sept. 16, 1886, in Case file 427, "Case File for the Queen v. *David J. Adams.*"
50. Examination of Charles Dankin, Captain of *Lansdowne*, September 17, 1886, and Examination of Frederick Allan, seaman of *Lansdowne*, coxswain, September 9, 1886, in Case file 427, "Case File for the Queen v. *David J. Adams.*"
51. "Article I, Convention of 20 October 1818, between the United States of America and His Britannic Majesty," in Joseph Pope, *A Short Account of the Negotiations Which Resulted in the Convention of the 20th October, 1818, between Great Britain and the United States, Together with the Text of the Treaty* (Ottawa, 1907).
52. Minutes of Filing, June 4, 1887, Argument of Defendants, by N. H. Meagher, in Case file 427,

"Case File for the Queen v. *David J. Adams*," 161.

53. Minutes of Filing, June 3, 1887, Argument of Wallace Graham, in Case file 427, "Case File for the Queen v. *David J. Adams*," 97.
54. Ibid., 104.
55. Ibid., 104.
56. Ibid., 106.
57. Ibid., 112.
58. In Minutes of Filing, June 4, 1887, Argument of Defendants, by N. H. Meagher, in Case file 427, "Case File for the Queen v. *David J. Adams*," 160.
59. "The David J. Adams: Confiscated for Violating the Fishery Treaty of 1818," *Digby Weekly Courier*, November 1, 1889.
60. "Letter from Phelps to Earl of Iddesleigh, September 11, 1886," in *Correspondence Related to the North American Fisheries*, 120–21.
61. "Letter from Thomas Bayard to Lionel Sackville-West, Department of State, Washington, May 10, 1886," in *Correspondence Related to the North American Fisheries: 1884–86, Presented to Both Houses by Command of Her Majesty, February 1887* (Printed by Her Majesty's Stationery Office, London, by Harrison & Sons, 1887), 38–40.
62. Letter from Mr. Phelps to the Earl of Rosebery, Legation of the United States, London, June 2, 1886, in *Correspondence Related to the North American Fisheries: 1884–86*, 50.
63. "Letter from Bayard to West, Department of State, Washington, May 29, 1886," in *Correspondence Related to the North American Fisheries: 1884–86*, 65; "Erring Sisters, Go in Peace," *New York Herald*, May 31, 1886.
64. "Hon. S. D. Thompson, Minister of Justice, Department of Justice, Report, 22 July 1886, in a Report of a Committee of the Honourable the Privy Council for Canada, Approved by His Excellency the Administrator of the Government in Council on the 2nd November, 1886, Included in Letter from Marquis of Lansdowne to Mr. Stanhope, Government House, Ottawa, 9 November 1886," *Correspondences Related to the North American Fisheries*, 170–84.
65. Clause 4: Construction of Subsequent Provisions of Act. Par II—Union. British North America Act. A Bill Entitled An Act for the Union of Canada, Nova Scotia, and New Brunswick, and the Government thereof; and for the Purposes Connected Therewith (Printed by the Earl of Carnarvon).
66. "Report of a Committee of the Honourable the Privy Council for Canada, Approved by his Excellency the Governor-General on the 14th June, 1886," *Correspondences Related to the North American Fisheries*, 81.
67. William P. Frye, "Fishing Rights of the United States, Speech of Hon. Wm. P. Frye, of Maine in the Senate of the United States, January 24, 1887," Maine Historical Society, O40 C685 V. 104:44, 3.
68. Frye, "Fishing Rights of the United States," 3.
69. Ibid., 5–6.
70. William Putnam, "Fishing Question between Great Britain and United States, April 16, 1888," in *The Book of the Fraternity Club, 1874–1924* (Portland, ME: Printed for the Club, 1925), 11.
71. Ibid., 21.
72. Ibid., 50–51; emphasis in original.
73. Ibid., 55–56.
74. As reported in "The Fisheries Dispute: The Yankees Looking After Their Coast Defense—The Retaliation Bill," *Novascotian*, February 5, 1887, 8.

75. "The Retaliation Bill," *Novascotian*, March 5, 1887, 2.
76. As quoted in "American Fishermen and Canadian Fisheries Regulations," *Novascotian*, July 23, 1887, 8.
77. "The Fisheries," *Novascotian*, May 7, 1887, 1.
78. "The Fisheries Question," *Novascotian*, May 2, 1887, 5.
79. "The Fisheries Question—Mr. Jones' Speech in the House of Commons," *Novascotian*, July 16, 1887, 3.
80. "Fishing Notes," *Novascotian*, May 28, 1887, 3. See also *Cape Ann Advertiser*, May 20, 1887.
81. "Seized for Smuggling," *Novascotian*, May 7, 1887, 6.
82. "Hiding Their Identity," *Novascotian*, June 18, 1887, 3.
83. "The Mackerel Fishery: Schools in Summerside Harbor—Americans Poaching without Molestation," *Novascotian*, July 9, 1887, 6.
84. "Two Seine Boats Caught—The Critic's Capture—Fourteen Men Made Prisoners at Souris," *Novascotian*, July 30, 1887, 6.
85. "The Shelburne Seizure—A Fine of $400 Imposed for Violating the Custom Law," *Novascotian*, July 30, 1887, 6.
86. "The Feeling in Boston—How the People Received the News of the Recent Seizures," *Novascotian*, August 6, 1887, 2.
87. U.S. Congress, Senate, Committee on Foreign Relations, *Report of the Committee on Foreign Relations in Relation to the Rights and Interests of American Fisheries and Fishermen* (Washington, DC: Government Printing Office, 1887), ix–x.
88. "Circular Issued by a Committee of the Boston Fish Bureau, Boston, September 1885, Included in a Letter from Lionel West to Marquis of Salisbury, October 10, 1885," *Correspondences Related to the North American Fisheries*, 27.
89. For a thorough history of American public opinion regarding the fisheries policy of Canada during the 1886–1888 period, see James Candow, "The North Atlantic Fisheries Dispute, 1886–1888, and Its Perception by the *New York Times* and the *New York Tribune*" (master's thesis, Dalhousie University, 1977).
90. William Frye, *Speech of Hon. Wm. P. Frye, of Maine in the Senate of the United States, January 24, 1887.*
91. U.S. Congress, Senate, Committee on Foreign Relations, *Report of the Committee on Foreign Relations in Relation to the Rights and Interests of American Fisheries and Fishermen*, 49th Cong., 2nd sess., 1886 (Washington, DC: Government Printing Office, 1887).
92. "The Fishery Question," *Novascotian*, September 10, 1887, 1.
93. Ibid., 2; "The Fishery Commission," *Novascotian*, October 22, 1887, 1.
94. *Record of the Proceedings of the Fisheries Conference at Washington, Nov. 1887*, reel 10; and *Other Printed Materials*, reels 10 and 11, Nova Scotia Archives and Records Management Microfilm 11,110.
95. U.S. Congress, Senate, Committee on Foreign Relations, *Views of the Majority, Report No. 3*, 50th Cong., 2nd sess., May 7, 1888 (Washington, DC: Government Printing Office, 1888).
96. U.S. Congress, Senate, Committee on Foreign Relations, *Views of the Minority, Report No. 3*, 50th Cong., 2nd sess., May 7, 1888 (Washington: Government Printing Office, 1888), 66–69.
97. Ibid., 89.
98. Ibid.
99. "Official Report of the Speech of Hon. Sir Charles Tupper, G.C.M.G., C.B., Minister of Finance of Canada, and One of Her Majesty's Plenipotentiaries at the Washington Fishery

Conference, on the Fishery Treaty, Delivered in the House of Commons of Canada, April 10th, 1888, Second Session, Sixth Parliament, 51 Vic.," in Sir Charles Tupper Papers, *Political Correspondences, under Memoranda and Report*, reel 10; and "Other Printed Materials," reels 10 and 11, Nova Scotia Archives and Records Management Microfilm 11,110.

100. "The Washington Treaty of 1888," *Novascotian*, February 25, 1888, 1.
101. Charles Stewart Davison, *Letters on the "Proposed" Fisheries Treaty of 1888* (New York: Published by Author, 1888); Joseph Doran, *Our Fishery Rights in the North Atlantic* (Philadelphia: Allen, Lane, and Scott's Printing House, 1888), 54.
102. Grover Cleveland, *Message from President to the 50th Congress, 1st Session* (Washington, DC: Government Printing Office, 1888).
103. Luther Maddocks, *Fisheries Treaties between the United States and Great Britain: Discussed from a Fishermen's Standpoint* (Washington, DC: Gray & Clarkson Printers, 1888), 1.
104. "Official Report of the Speech of Hon. Sir Charles Tupper, G.C.M.G., C.B., Minister of Finance of Canada, and One of Her Majesty's Plenipotentiaries at the Washington Fishery Conference, on the Fishery Treaty, Delivered in the House of Commons of Canada, April 10th, 1888."
105. "The Rejected Treaty: The Modus Vivendi Will Not Be Abrogated by Canada," *Novascotian*, August 25, 1888, 6.
106. "The Modus Vivendi," *Digby Weekly Courier*, June 1, 1888; "The Modus Vivendi," *Digby Weekly Courier*, April 19, 1889.
107. Stuart Weems Bruchey, *Enterprise: The Dynamic Economy of a Free Trade* (Cambridge, MA: Harvard University Press, 1990); Donald C. Masters, *Reciprocity, 1846–1911* (Ottawa: Canadian Historical Association, Booklet no. 12, 1961); Peter Morici, "Assessing the Canada-US Free Trade Agreement," *American Review of Canadian Studies* 26, no. 4 (Winter 1996): 491–98; Mildred A. Schwartz, "NAFTA and the Fragmentation of Canada," *American Review of Canadian Studies* 28, nos. 1–2 (Spring/Summer 1998): 11–28.
108. "The David J. Adams Case," *Digby Weekly Courier*, September 24, 1886.
109. Examination of George Vroom, of Clements, September 17, 1886, in Case file 427, "Case File for the Queen v. *David J. Adams*."
110. Examination of Robert Spurr, of Clements, September 17, 1886, in Case file 427, "Case File for the Queen v. *David J. Adams*."
111. Examination of Samuel D. Ellis, Victoria Beach, Sept. 17, 1886, in Case file 427, "Case File for the Queen v. *David J. Adams*."
112. The boarding record books for the Canadian Marine Police often cite "visiting relatives" as the reason for the presence of an American fishing schooner in Canadian waters. Most of these were subsequently warned off. This activity was also widely reported in the press; for example, see "The Captain of a Gloucester Fishermen Interviewed: Not Disposed to Give Up the Fish," *Digby Weekly Courier*, July 2, 1886, in which William Mclanson, captain of a Gloucester fishing schooner, noted that he was in St. Mary's Bay "to see my relatives" and that seven of his crew members were from his own home town of Weymouth, near Gilbert Cove, Nova Scotia. See also "Yankee Tricks," *Digby Weekly Courier*, October 12, 1888, which tells the story of an American crew stealing liquor from a store on shore while the captain was away visiting his wife's family.
113. "Seizure of the Gloucester Fisherman 'David J. Adams' for Violation of the Treaty," *Digby Weekly Courier*, May 14, 1886; "The Captured Fishermen," *Halifax Morning Herald*, May 10, 1886, 2.

114. "The Fisheries," *Digby Weekly Courier*, June 25, 1886.
115. "Protecting the Fisheries: A Talk with Capt. Scott, R.N.," *Digby Weekly Courier*, March 5, 1886.
116. "Cruise of the Lansdowne: American Fishermen Ordered off the Coast by Captain Scott," *Digby Weekly Courier*, April 2, 1886.
117. As quoted in "The Fisheries Question at Washington," *Novascotian*, December 12, 1885, 1.
118. "Cape Sable, Barrington, March 30," *Digby Weekly Courier*, April 2, 1886. Once the troubles of 1886 had passed, the American fishing fleet went back to the older method of picking up Canadian crews on their way to the fishing grounds. On March 8, 1889, the *Digby Weekly Courier* noted: "A dispatch from Halifax reports that advices from Shelburne county state that fishermen have received letters from American captains stating that they need not proceed to the United States this spring to obtain employment in the Yankee fishing fleet, as American vessels will call at Nova Scotia ports and ship men from their own homes."
119. "The Fisheries: Men Departing in Hundreds for the States," *Novascotian*, April 3, 1886, 8.
120. "Off to Join Their Vessels," *Halifax Chronicle*, March 29, 1886.
121. "The Fisheries: The Fishery Warrior on the Shelburne Coast," *Novascotian*, April 3, 1886, 8.
122. As quoted in "Canadian Fisheries—Attitudes of Our Government Not Approved by the Americans," *Digby Weekly Courier*, April 9, 1886.
123. "Falling off of American Fishermen," *Digby Weekly Courier*, June 18, 1886.
124. "American Fishing Note," *Digby Weekly Courier*, March 8, 1889.
125. For a critique of the nationalism and overzealous pride in Canada and the United States, see "The Fishery Negotiations," *Novascotian*, February 4, 1888, 1.

4. "PEACEABLE SETTLEMENT": BAIT AND INTERNATIONAL LAW, 1888–1910

1. C. P. Stacey, *Canada and the Age of Conflict*, vol. 1, *1867–1921* (Toronto: University of Toronto Press, 1984), 37, 39–40.
2. U.S. Commission of Fish and Fisheries, *Part XII, Report of the Commission for 1886* (Washington, DC: Government Printing Office, 1889), xi–xii.
3. Frederick Rowe, *A History of Newfoundland and Labrador* (Toronto: McGraw-Hill Ryerson, 1980), 298–99, 323–24.
4. Ibid., 327–68.
5. Ibid., 303–6.
6. Ibid., 302.
7. Ibid., 346.
8. Raymond B. Blake, *From Fishermen to Fish: The Evolution of Canadian Fishery Policy* (Toronto: Irwin Press, 2000), 22–23.
9. Herring, caplin, and squid were all typical bait species for the Grand Banks's cod fisheries.
10. As quoted in North Atlantic Coast Fisheries Arbitration, *The Case of the United States before the Permanent Court of Arbitration at The Hague, under the Provisions of the Special Agreement between the United States of America and Great Britain Concluded January 27, 1909*, vol. 1 of *Proceedings in the North Atlantic Coast Fisheries Arbitration before the Permanent Court of Arbitration at The Hague, under the Provisions of the General Treaty of Arbitration of April 4, 1908, and the Special Agreement of January 27, 1909, between the United States of America and Great Britain* (Washington, DC: Government Printing Office, 1912), 226.
11. Ibid., 208.

12. C. P. Stacy, *Canada and the Age of Conflict*, 85–113.
13. Ibid., 244.
14. Frederick Rowe, *A History of Newfoundland and Labrador*, 306.
15. Elihu Root, Robert Bacon, and James Brown Scott, eds., *The North Atlantic Coast Fisheries Arbitration at The Hague: Argument on Behalf of the United States* (Cambridge, MA: Harvard University Press, 1917), 23–24.
16. Elihu Root, *Latin America and the United States* (Cambridge, MA: Harvard University Press, 1917), 230–31.
17. North Atlantic Coast Fisheries Arbitration, *The Case of the United States before the Permanent Court of Arbitration at The Hague*, 37–38.
18. James Brown Scott, ed., *Argument of the Honorable Elihu Root on Behalf of the United States: Before the North Atlantic Coast Fisheries Arbitration Tribunal at the Hague, 1910* (The World Peace Foundation, 1912), iii.
19. Root, Bacon, and Scott, eds., *The North Atlantic Coast Fisheries Arbitration at The Hague*, lxxvii–lxxxi.
20. North Atlantic Coast Fisheries Arbitration, *The Case of Great Britain before the Permanent Court of Arbitration at The Hague, under the Provisions of the Special Agreement between the United States of America and Great Britain Concluded January 27, 1909, with Parts I & II of the Appendix to the Case of Great Britain Including Treaties and Correspondences*, vol. 4 of *Proceedings in the North Atlantic Coast Fisheries Arbitration before the Permanent Court of Arbitration at The Hague*, 42–43.
21. North Atlantic Coast Fisheries Arbitration, *The Case of the United States before the Permanent Court of Arbitration at The Hague*, 222.
22. North Atlantic Coast Fisheries Arbitration, *The Case of Great Britain before the Permanent Court of Arbitration at The Hague*, 20–21, 45–48.
23. North Atlantic Coast Fisheries Arbitration, *The Counter Case of Great Britain before the Permanent Court of Arbitration at The Hague, under the Provisions of the Special Agreement between the United States of America and Great Britain Concluded January 27, 1909*, vol. 7 of *Proceedings in the North Atlantic Coast Fisheries Arbitration before the Permanent Court of Arbitration at The Hague*, 8.
24. U.S. Commission of Fish and Fisheries, *Part I, Report on the Condition of the Sea Fisheries of the South Coast of New England in 1871 and 1872* (Washington, DC: Government Printing Office, 1873), x–xiii.
25. North Atlantic Coast Fisheries Arbitration, *The Case of the United States before the Permanent Court of Arbitration at The Hague*, 63.
26. Ibid., 10, 37–38.
27. Ibid., 71–76.
28. Ibid., 95.
29. Ibid., 98.
30. Ibid., 130–46.
31. Ibid., 162.
32. Ibid., 189–90.
33. Ibid., 206.
34. Ibid., 207.
35. Ibid., 224.
36. "Letter from Root to Grey, 22 January 1906," as quoted in North Atlantic Coast Fisheries

Arbitration, *The Case of the United States before the Permanent Court of Arbitration at The Hague*, 225.

37. North Atlantic Coast Fisheries Arbitration, *The Case of Great Britain before the Permanent Court of Arbitration at The Hague*, 22.
38. "Letter from Grey to Root, 2 February 1906," as quoted in North Atlantic Coast Fisheries Arbitration, *The Case of the United States before the Permanent Court of Arbitration at The Hague*, 215.
39. North Atlantic Coast Fisheries Arbitration, *The Case of Great Britain before the Permanent Court of Arbitration at The Hague*, 25.
40. Ibid., 20.
41. Ibid., 41.
42. Blake, *From Fishermen to Fish: The Evolution of Canadian Fishery Policy*. See also Susan Hanna, Heather Blough, Richard Allen et al., *Fishing Grounds: Defining a New Era for American Fisheries Management* (Washington, DC: Island Press, 2000), ix–x.

CONCLUSION

1. Sean Cadigan, "The Moral Economy of the Commons: Ecology and the Equity in Newfoundland Cod Fisheries, 1815–1855," *Labour/Le Travail* 43 (Spring 1999): 9–42.
2. Some work has been done on this. See Archie MacLean, *The Problems of the Fishing Industry of Eastern Nova Scotia: Fieldwork in Fisheries for St. Francis Xavier Extension Department* (Antigonish, NS: September 1961, Nova Scotia Archives and Records Management, Manuscript Collection); L. Gene Barrett, "Development and Underdevelopment, and the Rise of Trade Unionism in the Fishing Industry of Nova Scotia, 1900–1950," (master's thesis, Dalhousie University, 1976); William Ernst, MP, *Submission on Behalf of the Fishermen of the Province of Nova Scotia as Presented to the Royal Commission on Fisheries at Cheticamp, NS* (Published by Authority of Hpn. J.A. Walker, Minister of Natural Resources, c. 1929).
3. "Forty-Second Annual Report of the Department of Marine and Fisheries, 1908–9, Fisheries," *Sessional Papers*, 1910, no. 22, xxi.
4. Ibid., xxii–xxiii.
5. William Wakeham, "Appendix No. 6—Province of Quebec," in "Forty-Second Annual Report of the Department of Marine and Fisheries, 1908–9, Fisheries," *Sessional Papers*, 1910, no. 22, 157.
6. John J. Bowie, "Appendix No. 20—Steam Trawling—Beam and Otter," in "Forty-Second Annual Report of the Department of Marine and Fisheries, 1908–9, Fisheries," *Sessional Papers*, 1910, no. 22, 373–85.
7. See Richard Judd, "Grass-Roots Conservation in Eastern Coastal Maine: Monopoly and the Moral Economy of Weir Fishing, 1893–1911," *Environmental Review* 12, no. 2 (Summer 1988): 81–103; and "Saving the Fishermen as Well as the Fish: Conservation and Commercial Rivalry in Maine's Lobster Industry: 1872–1933," *Business History Review* 62 (1988): 596–625.

Bibliography

Abreu-Ferreira, Darlene. "Portugal's Cod Fishery in the 16th Century: Myths and Misconceptions." In *How Deep is the Ocean? Historical Essays on Canada's Atlantic Fishery*, edited by James Candow and Carol Corbin. Sydney, NS: University College of Cape Breton Press, 1997.

Acheson, James. *Capturing the Commons: Devising Institutions to Manage the Maine Lobster Industry.* Hanover, NH: University Press of New England, 2003.

———. *The Lobster Gangs of Maine.* Hanover, NH: University Press of New England, 1988.

Alexander, David. "New Notions of Happiness: Nationalism, Regionalism and Atlantic Canada." *Journal of Canadian Studies* 15 (Summer 1980): 29–42.

Andersen, Raul. "Nineteenth-Century American Banks Fishing under Its Health and Injury Costs." *Canadian Folklore Canadien* 12, no. 2 (1990): 101–22.

Ankli, R. E. "Canadian-American Reciprocity: A Comment." *Journal of Economic History* 28 (June 1970): 274–81.

———. "The Reciprocity Treaty of 1854." *Canadian Journal of Economic History* 4 (1971): 1–20.

Armytage, Lowell. *The Free Port System in the British West Indies: A Study in Commercial Policy, 1766–1833.* New York: Longmans, Green, and Co., 1953.

Balcom, Berton A. *History of the Lunenburg Fishing Industry.* Lunenburg, NS: Lunenburg Marine Museum Society, 1977.

Barrett, Gene. "Development and Underdevelopment, and the Rise of Trade Unionism in the Fishing Industry of Nova Scotia, 1900–1950." Master's thesis, Dalhousie University, 1976.

———. "Underdevelopment and Social Movements in the Nova Scotia Fishing Industry to 1938." In *Underdevelopment and Social Movements in Atlantic Canada*, edited by Robert J. Brym and R. James Sacouman. Toronto: Hogtown Press, 1976.

Battick, John F. "A Survey of Primary Sources for the Social and Economic History of Seafaring Communities in Maine." *Maine Historical Society Quarterly* 24 (1985): 394–400.

Bayard, Thomas. *Selected Cases of Maltreatment of American Fishing Vessels.* Washington, DC: Government Printing Office, 1887.

Beck, J. Murray. *Politics of Nova Scotia.* Vol. 1, *Nicholson-Fielding, 1710–1896.* Tantallon, NS: Four East Publications, 1985.

Bière, Jean-François. "The French Fishery in North America in the 18th Century." In *How Deep is the Ocean? Historical Essays on Canada's Atlantic Fishery,* edited by James Candow and Carol Corbin. Sydney, NS: University College of Cape Breton Press, 1997.

Benns, F. Lee. *The American Struggle for the British West Indian Trade, 1815–1830.* Bloomington: University of Indiana Press, 1923.

Blake, Raymond B. *From Fishermen to Fish: The Evolution of Canadian Fishery Policy.* Toronto: Irwin Press, 2000.

Brebner, John B. *North Atlantic Triangle: The Interplay of Canada, the United States, and Great Britain.* 2nd ed. Toronto: Ryerson Press, 1964.

Brookes, Alan. "Out-Migration from the Maritime Provinces, 1860–1900: Some Preliminary Considerations." *Acadiensis* 5 (Spring 1976): 26–55.

Bruchey, Stuart Weems. *Enterprise: The Dynamic Economy of a Free Trade.* Cambridge, MA: Harvard University Press, 1990.

———. *Growth of the Modern American Economy.* New York: Dodd, Mead & Company, 1975.

Buckner, Phillip A. *The Transition to Responsible Government: British Policy in British North America, 1815–1850.* Westport, CT: Greenwood Press, 1985.

———. "The 1870s: Political Integration." In *The Atlantic Provinces in Confederation,* edited by E. R. Forbes and D. A. Muise. Toronto: University of Toronto Press, 1993.

Buckner, Phillip A., and David Frank, eds. *Acadiensis Reader.* Vol. 1, *Atlantic Canada before Confederation.* Fredericton, NB: Acadiensis Press, 1990.

———, eds. *Acadiensis Reader.* Vol. 2, *Atlantic Canada after Confederation.* Fredericton, NB: Acadiensis Press, 1988.

Buckner, Phillip A., and John G. Reid, eds. *The Atlantic Region to Confederation: A History.* Toronto: University of Toronto Press, 1994.

Burrill, Gary. *Away: Maritimers in Massachusetts, Ontario, and Alberta.* Montreal: McGill-Queen's Press, 1992.

Cadigan, Sean. "Failed Proposals for Fisheries Management and Conservation in Newfoundland, 1855–1880." In *Fishing Places, Fishing People: Issues in Small Scale Fisheries,* edited by Dianne Newell and Rosemary Ommer. Toronto: University of Toronto Press, 1998.

———. *Hope and Deception in Conception Bay: Merchant-Settler Relations in Newfoundland, 1785–1855.* Toronto: University of Toronto Press, 1995.

———. "The Moral Economy of the Commons: Ecology and the Equity in Newfoundland Cod Fisheries, 1815–1855." *Labour/Le Travail* 43 (Spring 1999): 9–42.

Calhoun, Charles, ed. *The Gilded Age: Perspectives on the Origins of Modern America.* 2nd ed. Lanham, MD: Rowman & Littlefield, 2007.

Campbell, Charles, *From Revolution to Rapprochement: The United States and Great Britain, 1783–1900.* New York: Wiley, 1974.

Candow, James. "The North Atlantic Fisheries Dispute, 1886–1888, and Its Perception by the New York Times and the New York Tribune." Master's thesis, Dalhousie University, 1977.

Candow, James, and Carol Corbin, eds. *How Deep Is the Ocean? Historical Essays on Canada's Atlantic*

Fishery. Sydney, NS: University College of Cape Breton Press, 1997.
Chamberlyne, Charles F. *State Rights in State Fisheries: An Argument by Charles F. Chamberlyne, Esq. of Boston, Mass., as Counsel for the Commission of Sea and Shore Fisheries of Maine, before the Committee on Merchant Marine and Fisheries, February 24, 1892 in Opposition to the "Lapham Bill" to Permit Seining for Mackerel and Menhaden in State Waters Contrary to State Law.* Washington, DC: Rufus H. Darby and Job Printer, 1892.
Chapelle, Howard I. *The American Fishing Schooners, 1825–1935.* New York: W.W. Norton & Co., 1973.
Clark, Andrew Hill. *Three Centuries and the Island: A Historical Geography of Settlement and Agriculture in Prince Edward Island, Canada.* Toronto: University of Toronto Press, 1959.
Clark, Lovell. "Regionalism? Or Irrationalism?" *Journal of Canadian Studies* 13 (Summer 1978): 119–24.
Cook, Ramsey. "The Burden of Regionalism." *Acadiensis* 7 (Autumn 1977): 110–15.
———. "Regionalism Unmasked." *Acadiensis* 13 (Autumn 1983): 137–42.
Crowley, John E. "Empire versus Trunk: The Official Interpretation of Debt and Labour in the Eighteenth-Century Newfoundland Fishery." *Canadian Historical Review* 70, no. 3 (September 1989): 311–36.
Davison, Charles Stewart. *Letters on the "Proposed" Fisheries Treaty of 1888.* New York: Published by Author, 1888.
DeRicci, James H. *The Fisheries Dispute and Annexation of Canada.* London: Marston, Searle & Rivington Ltd., 1888.
Doran, Joseph. *Our Fishery Rights in the North Atlantic.* Philadelphia: Allen, Lane, and Scott's Printing House, 1888.
DuBoff, Richard. *Accumulation and Power: An Economic History of the United States.* Armonk, NY: ME Sharpe, 1989.
Duncan, Roger, F. *Coastal Maine: A Maritime History.* New York: W.W. Norton Press, 1992.
Eldridge, C. C. *England's Mission: The Imperial Idea in the Age of Gladstone and Disraeli, 1868–1880.* New York: Macmillan Press, 1973.
Ernst, William M. P. Submission on Behalf of the Fishermen of the Province of Nova Scotia as Presented to the Royal Commission on Fisheries at Cheticamp, NS. Published by Authority of Hon. J. A. Walker, Minister of Natural Resources, 1929.
Fieldhouse, D. K. *Economics and Empires, 1830–1914.* Ithaca, NY: Cornell University Press, 1973.
Fingard, Judith. "The 1880s: Paradoxes of Progress." In *The Atlantic Provinces in Confederation*, edited by E. R. Forbes and D. A. Muise. Toronto: University of Toronto Press, 1993.
Forbes, E. R., and D. A. Muise, eds. *The Atlantic Provinces in Confederation.* Toronto: University of Toronto Press, 1993.
Freehling, William W. *The Road to Disunion: Secession at Bay, 1776–1854.* New York: Oxford University Press, 1990.
Gallagher, John, and Ronald Robinson. "The Imperialism of Free Trade." *Economic History Review*, 2nd series, 6, no. 1 (1953): 1–15.
Garland, Joseph. *The Fishing Schooners of Gloucester.* Boston: DR Godine, 1983.
Gerriets, Marilyn, and Julian Gwyn. "Tariffs, Trade, and Reciprocity: Nova Scotia 1830–1866." *Acadiensis* 25, no. 2 (Spring 1996): 62–81.
Golladay, V. Dennis. "The United States and British North American Fisheries, 1815–1818." *American Neptune* 33, no. 4 (1973): 246–57.
Goode, George Brown. *The Fisheries and Fishing Industry of the United States.* 5 vols. Washington,

DC: Government Printing Office, 1884–1887.

Gordon, H. Scott. "The Economic Theory of a Common-Property Resource: The Fishery." *Journal of Political Economy* 62 (1954): 124–42.

Gough, Joseph. *Fisheries Management in Canada, 1880–1910*. Ottawa: Canadian Manuscript Report of Fisheries and Aquatic Sciences, Supply and Services of Canada, 1991.

Gould, Lewis L. "Party Conflict: Republicans versus Democrats, 1877–1910." In *The Gilded Age: Perspectives on the Origins of Modern America*, 2nd ed., edited by Charles Calhoun. Lanham, MD: Rowman & Littlefield Publishers, 2007.

Great Britain. *Correspondence with the Government of Canada in Connection with the Appointment of the Joint Commission and the Treaty of Washington: Presented to Both Houses of Parliament by Command of Her Majesty, April, 1872*. London: W. Clowes & Son, for Her Majesty's Stationery Office, 1872.

———. *Correspondences Related to the North American Fisheries, 1884–86*. London: Printed by Her Majesty's Stationery Office, Harrison & Son, 1887.

Greer, Allan, and Ian Ranforth, eds. *Colonial Leviathan: State Formation in Mid-Nineteenth-Century Canada*. Toronto: University of Toronto Press, 1992.

Grenville, J. A. S. *Lord Salisbury and Foreign Policy: The Close of the Nineteenth Century*. London: University of London Press, 1964.

Grettler, David J. "The Nature of Capitalism: Environmental Change and Conflict over Commercial Fishing in Nineteenth-Century Delaware." *Environmental History* 6 (July 2001): 451–73.

Gwyn, Julian, "Golden Age or Bronze Age? Wealth and Poverty in Nova Scotia: The 1850s and 1860s." In *Canadian Papers in Rural History*, edited by Donald H. Akensen. Gananoque, ON: Langdale Press, 1992.

———. "Tariffs, Trade, and Reciprocity: Nova Scotia 1830–1866." *Acadiensis* 25, no. 2 (Spring 1996): 62–81.

Halifax Commission. *Records of the Proceedings of the Halifax Commission, 1877 under the Treaty of Washington of May 8, 1871*. Washington, DC: Government Printing Office, 1878.

Hanna, Susan, Heather Blough, Richard Allen, Suzanna Iudicell, Gary Matlock, and Bonnie McCay. *Fishing Grounds: Defining a New Era for American Fisheries Management*. Washington, DC: Island Press, 2000.

Hardin, Garrett. "The Tragedy of the Commons." *Science* 162 (1968): 1243–48.

Hardin, Garrett, and John Baden. *Managing the Commons*. San Francisco: W. H. Freeman, 1977.

Hecht, Irene. "Israel D. Andrews and the Reciprocity Treaty of 1854: A Reappraisal." *Canadian Historical Review* 44 (December 1963): 313–26.

Hiller, J. K. "The Newfoundland Fisheries Issue in Anglo-French Treaties, 1713–1904." *Journal of Imperial and Commonwealth History* 24, no. 1 (January 1996): 1–23.

Hilton, George W. *The Truck System Including a History of the British Truck Acts, 1465–1960*. Westport, CT: W. Heffer, 1960.

Hornsby, Stephen J. *British Atlantic, American Frontier: Space of Power in Early Modern British America*. Hanover, NH: University of New England Press, 2005.

Hornsby, Stephen J., Victor A. Konrad, and James J. Herlan, eds. *The Northeastern Borderlands: Four Centuries of Interaction*. Fredericton, NB: Acadiensis Press, 1989.

Hubbard, Jennifer. *A Science on the Scale: The Rise of Canadian Fisheries Biology, 1898–1939*. Toronto: University of Toronto Press, 2006.

Innis, Harold. *The Cod Fisheries: The History of an International Economy*. New Haven, CT: Yale University Press, 1940.

Isham, Charles. *The Fishery Question: Its Origin, History, & Present Situation: With a Map of the Anglo-American Fishing Grounds & a Short Bibliography*. New York: Putnam's Sons, 1887.

Jones, Howard. *To the Webster-Ashburton Treaty: A Study in Anglo-American Relations, 1783–1843*. Ann Arbor: University of Michigan Press, 1973.

———. *Union in Peril: The Crisis over British Intervention in the Civil War*. Chapel Hill: University of North Carolina, 1992.

Jones, Wilbur D. *The American Problem in British Diplomacy, 1841–1861*. Athens: University of Georgia Press, 1974.

Judd, Richard. *Aroostook: A Century of Logging in Northern Maine*. Orono, ME: University of Maine Press, 1988.

———. *Common Lands, Common People: The Origins of Conservation in Northern New England*. Cambridge, MA: Harvard University Press, 1997.

———. "Grass-Roots Conservation in Eastern Coastal Maine: Monopoly and the Moral Economy of Weir Fishing, 1893–1911." *Environmental Review* 12, no. 2 (Summer 1988): 81–103.

———. "Saving the Fishermen as Well as the Fish: Conservation and Commercial Rivalry in Maine's Lobster Industry: 1872–1933." *Business History Review* 62 (1988): 596–625.

Kilby, William Henry. *Eastport and Passamaquoddy: A Collection of Historical and Bibliographic Sketches*. Eastport, ME: Edward E. Shead & Co., 1988.

Konrad, Victor. *Borderlands: Essays in Canadian-American Relations*. Toronto: E.C.W. Press, 1991.

———. "The Borderlands of the United States and Canada in the Context of North American Development." *International Journal of Canadian Studies* 4 (Fall 1991): 77–96.

Kurlansky, Mark. *Cod: Biography of the Fish That Changed the World*. New York: Penguin Press, 1997.

LaBaree, Benjamin W. *America and the Sea: A Maritime History*. Mystic, CT: Mystic Seaport Publications, 1998.

Lee, David. *The Robbins of Gaspé, 1766–1825*. Markham, ON: Fitzhenry & Whiteside, 1984.

Littlefield, Charles E. *Fisheries within the Territorial Limits of the State Are Not Subject to Congressional Control. A Reply on Behalf of the State of Maine to the Argument Submitted to Sustain the Lapham Bill (H.R. 5030) before the Congressional Committee of Merchant Marine and Fisheries of the 52nd Congress*. Boston: Press of Rockwell & Churchill, 1892.

MacCany, R. C. "Buyer Concentration: The Inshore Groundfish Processing Industry in Nova Scotia." *Canadian Journal of Regional Science* 12, no. 2 (Summer 1989): 247–58.

MacDonald, David A. "They Cannot Pay Us in Money: Newman and Company and the Supplying System in the Newfoundland Fishery, 1850–1884." *Acadiensis* 19, no. 1 (Fall 1989): 142–56.

MacLean, Archie. *The Problems of the Fishing Industry of Eastern Nova Scotia: Fieldwork in Fisheries for St. Francis Xavier Extension Department*. Antigonish, NS: September 1961. Nova Scotia Archives and Records Management, Manuscript Collection.

Maddocks, Luther. *Fisheries Treaties between the United States and Great Britain: Discussed from a Fisherman's Standpoint*. Washington: Gray & Clarkson Printers, 1888.

Masters, Donald C. *Reciprocity, 1846–1911*. Ottawa: Canadian Historical Association, Booklet no. 12, 1961.

McCay, Bonnie, J., and James Acheson, eds. *The Question of the Commons: The Culture and Ecology of Communal Resources*. Tucson: The University of Arizona Press, 1987.

McDonald, R. H. "Nova Scotia and the Reciprocity Negotiations, 1845–1854: A Re-Interpretation." *Nova Scotia Historical Quarterly* (1977): 205–34.

McEvory, Arthur F. *The Fisherman's Problem: Ecology and Law in California Fisheries, 1850–1980*. New York: Cambridge University Press, 1986.

McKinsey, Lauren, and Victor Konrad. *Borderlands Reflections: The United States and Canada.* Orono: University of Maine Press, 1989.

Morici, Peter. "Assessing the Canada-US Free Trade Agreement." *American Review of Canadian Studies* 26, no. 4 (Winter 1996).

Morison, Samuel Eliot. *The Maritime History of Massachusetts Bay, 1783–1860.* Boston: Houghton Mifflin Company, 1921.

Mowat, Farley. *Sea of Slaughter.* Toronto: McClelland and Stewart, 1984.

Murdock, Richard. "Cod or Mackerel: Bounty Payment Disputes, 1829–32." *Essex Institute Historical Collections* 105 (October 1969): 309–15.

North Atlantic Coast Fisheries Arbitration. *Proceedings in the North Atlantic Coast Fisheries Arbitration before the Permanent Court of Arbitration at The Hague, under the Provisions of the General Treaty of Arbitration of April 4, 1908, and the Special Agreement of January 27, 1909, between the United States of America and Great Britain.* Vols. 1–10. Washington, DC: Government Printing Office, 1912.

Nova Scotia Department of Fisheries. *Sea, Salt, and Sweat: A Short Story of Nova Scotia and the Vast Atlantic Fishery.* Halifax: Nova Scotia Department of Fisheries, 1996.

O'Leary, Wayne. *Maine Sea Fisheries: The Rise and Fall of a Native Industry.* Boston: Northeastern University, 1996.

Officer, L., and L. Smith. "The Canadian American Reciprocity Treaty of 1855 to 1868." *Journal of Economic History* 28 (1968): 598–623.

Ommer, Rosemary. *From Outpost to Outpost: A Structural Analysis of the Jersey-Gaspé Cod Fishery, 1767–1886.* Montreal: McGill-Queen's University Press, 1991.

———, ed. *Merchant Credit and Labour Strategies in Historical Perspective.* Fredericton, NB: Acadiensis Press, 1990.

———. "One Hundred Years of Fishery Crisis in Newfoundland." *Acadiensis* 23, no. 2 (Spring 1994): 5–20.

———. "The Trunk System in Gaspé, 1822–77." *Acadiensis* 19, no. 1 (Autumn 1988): 91–115.

Painter, Nell Irvin. *Standing at Armageddon: The United States, 1877–1919.* New York: W.W. Norton & Company, 1987.

Pearl, Cephas. *Journal of Cephas Pearl, Born May 4, 1863 at Tancook, Sailed on Various Ships in 1880s.* Nova Scotia Archives and Records Management, MG 7: vol. 14a.

Perkins, Bradford. *Castlereagh and Adams; England and the United States, 1812–1823.* Berkeley: University of California Press, 1964.

Perley, Moses H. *Report on the Seas and River Fisheries of New Brunswick, Laid before the House of the Assembly by Command of His Excellency the Lieutenant Governor.* Fredericton, NB: J. Simpson, Printer to the Queen's Most Excellent Majesty, 1852.

———. *Report upon the Fisheries of the Bay of Fundy, Laid before the House of the Assembly by Command of His Excellency the Lieutenant Governor, March 15, 1851.* Fredericton, NB: J. Simpson, Printer to the Queen's Most Excellent Majesty, 1851.

Pope, Joseph. *A Short Account of the Negotiations Which Resulted in the Convention of the 20th October, 1818, between Great Britain and the United States, Together with the Text of the Treaty.* Ottawa, 1907.

Pope, Peter. *Fish into Wine: The Newfoundland Plantation in the Seventeenth Century.* Chapel Hill: University of North Carolina Press, 2004.

———. "The 16th-Century Fishing Voyage." In *How Deep is the Ocean? Historical Essays on Canada's Atlantic Fishery*, edited by James Candow and Carol Corbin. Sydney, NS: University College

of Cape Breton Press, 1997.
Putnam, William. "Fishing Question between Great Britain and United States, April 16, 1888." In *The Book of the Fraternity Club, 1874–1924*. Portland, ME: Printed for the Club, 1925.
Ramsay, R. A. *Treaties Affecting the Boundaries and Fisheries of Canada.* Montreal: Young Men's Association of St. Paul's Church, 1885.
Raymond, Gregory A. "Canada between the Superpowers: Reciprocity and Conformity in Foreign Policy." *American Review of Canadian Studies* 17 (Summer 1987): 221–36.
Root, Elihu. *Latin America and the United States.* Cambridge, MA: Harvard University Press, 1917.
Root, Elihu, Robert Bacon, and James Brown Scott, eds. *The North Atlantic Coast Fisheries Arbitration at The Hague: Argument on Behalf on the United States.* Cambridge, MA: Harvard University Press, 1917.
Rothney, G. O. "British Policy in the North American Cod-Fisheries, with Special Reference to Foreign Competition, 1775–1819." Ph.D. diss., University of London, 1939.
Rowe, Frederick. *A History of Newfoundland and Labrador.* Toronto: McGraw-Hill Ryerson, 1980.
Ryan, Shannon. "Fishery to Colony: A Newfoundland Watershed, 1793–1815." In *The Acadiensis Reader*, vol. 1, 3rd ed., edited by P. A. Buckner, Gail G. Campbell, and David Frank. Fredericton, NB: The Acadiensis Press, 1998.
Samson, Roch. *Fishermen and Merchants in Nineteenth Century Gaspé: The Fishermen-Dealers of William Hyman and Sons.* Ottawa: National Historic Parks and Site Branch, Parks Canada, 1984.
Saunders, S. A. "The Maritime Provinces and the Reciprocity Treaty." *Dalhousie Review* 14 (October 1934): 22–43.
———. "The Reciprocity Treaty of 1854: A Regional Study." *Canadian Journal of Economic and Political Science* 2 (February 1936): 41–53.
Schwartz, Mildred A. "NAFTA and the Fragmentation of Canada." *American Review of Canadian Studies* 28, nos. 1–2 (Spring/Summer 1998): 11–28.
Scott, James Brown, ed. *Argument of the Honorable Elihu Root on Behalf of the United States: Before the North Atlantic Coast Fisheries Arbitration Tribunal at The Hague, 1910.* Washington, DC: The World Peace Foundation, 1912.
See, Scott W. *The History of Canada.* Westport, CT: Greenwood Press, 2001.
Setser, Vernon G. *The Commercial Reciprocity Policy of the United States, 1774–1829.* New York: Da Capo Press, 1937.
Stacey, C. P. *Canada and the Age of Conflict.* Vol. 1, *1867–1921.* Toronto: University of Toronto Press, 1984.
Steinberg, Theodore. *Nature Incorporated: Industrialization and the Waters of New England.* New York: Cambridge University Press, 1991.
Stevenson, Garth. "Canadian Regionalism in Continental Perspective." *Journal of Canadian Studies* 15, no. 2 (Summer 1980): 25–51.
Stewart, Alice R. "A Regional Approach for New England and the Atlantic Provinces." *Canadian Historical Association, Annual Report* (1966): 187–91.
Sturgis, James. "The Opposition to Confederation in Nova Scotia, 1864–1868." In *The Causes of Canadian Confederation*, edited by Ged Martin. Fredericton, NB: Acadiensis Press, 1990.
Swainger, Jonathan. *The Canadian Department of Justice and the Completion of Confederation, 1867–78.* Vancouver: University of British Columbia Press, 2000.
Tallman, Ronald D. "'Annexation in the Maritimes?' The Butler Mission to Charlottetown." *Dalhousie Review* 53 (Spring 1973): 97–112.
———. "Warships and Mackerel: The North Atlantic Fisheries in Canadian-American Relations,

1867–1877." Ph.D. diss., University of Maine, 1971.

Tennyson, Brian D. "Economic Nationalism, Confederation and Nova Scotia." In *The Causes of Canadian Confederation*, edited by Ged Martin. Fredericton, NB: Acadiensis Press, 1990.

Thompson, John Herd, and Stephen J. Randall. *Canada and the United States: Ambivalent Allies.* 2nd ed. Athens: University of Georgia Press, 1997.

Tulloch, Judith. "The New England Fishery and Trade at Canso, 1720–1744." In *How Deep is the Ocean? Historical Essays on Canada's Atlantic Fishery*, edited by James Candow and Carol Corbin. Sydney, NS: University College of Cape Breton Press, 1997.

U.S. Commission of Fish and Fisheries. *Report of the Commission.* Washington, DC: Government Printing Office, 1873–1892.

Vaughan, Frederick. *The Canadian Federalist Experiment: From Defiant Monarch to Reluctant Republic.* Montreal: McGill-Queen's Press, 2003.

Vickers, Daniel. *Farmers and Fishermen: Two Centuries of Work in Essex County, Massachusetts, 1630–1850.* Chapel Hill: University of North Carolina Press, 1994.

———, ed. *Marine Resources and Human Societies in the North Atlantic since 1500.* St. John's, NF: Memorial University of Newfoundland, 1995.

———. "The Price of Cod: A Price Index for Cod, 1505–1892." *Acadiensis* 25, no. 2 (Spring 1996): 92–104.

Waite, P. B. *The Life and Times of Confederation, 1864–1867: Politics, Newspapers and the Union of British North America.* 3rd ed. Toronto: University of Toronto Press, 1962.

Wallace, Frederick William. *Roving Fisherman: An Autobiography, Recounting Personal Experience in the Commercial Fishing Fleets and Fish Industry of Canada and the United States, 1911–1924.* Gardenvale, QC: Canadian Fishermen, 1955.

Ward, John Manning. *Colonial Self-Government: The British Experience, 1759–1856.* Toronto: University of Toronto Press, 1976.

Webb, Paul. "British Squadrons in North American Waters, 1783–1797." *Northern Mariner* 5, no. 2 (April 1995): 1–18.

Weber, Michael L. *From Abundance to Scarcity: A History of U.S. Marine Fisheries Policy.* Washington, DC: Island Press, 2002.

Wells, Kennedy. *The Fisheries of Prince Edward's Island.* Charlottetown, PEI: Ragweed Press, 1986.

White, Patrick C. T. *The Critical Years: American Foreign Policy, 1793–1823.* New York: Wiley, 1970.

Winsor, Frederick Archibald. "The Newfoundland Bank Fishery: Government Policies and the Struggle to Improve Bank Fishing Crews' Working, Health, and Safety Conditions, 1876–1920." Ph.D. diss., Memorial University of Newfoundland, 1997.

Woodbury, Charles Levi. *The Relations of the Fisheries to the Discovery and Settlement of North America; Delivered before the New Hampshire Historical Society, at Concord, June, 1880, and the Massachusetts Fish and Game Protection Society, at Boston, 1880.* Boston: Alfred Mudge & Son, Printers, 1880.

Woodward, C. Vann. *The Comparative Approach to American History.* New York: Basic Books, 1968.

Worster, Donald. *Dust Bowl: The Southern Plains in the 1930s.* New York: Oxford University Press, 1979.

Index